AVRO
SHACKLETON

1949 to 1991 (all marks)

COVER CUTAWAY: Avro Shackleton MR.3.

(Mike Badrocke)

First published in September 2015

Keith Wilson has asserted his moral right to be identified as the author of this work.

A catalogue record for this book is available from the British Library.

ISBN 978 0 85733 769 6

Library of Congress control no. 2014956837

Published by Haynes Publishing,
Sparkford, Yeovil,
Somerset BA22 7JJ, UK.
Tel: 01963 440635
Int. tel: +44 1963 440635
Website: www.haynes.co.uk

Haynes North America Inc.,
861 Lawrence Drive, Newbury Park,
California 91320, USA.

Printed in the USA by
Odcombe Press LP,
1299 Bridgestone Parkway,
La Vergne, TN 37086.

AVRO
SHACKLETON

1949 to 1991 (all marks)

Owners' Workshop Manual

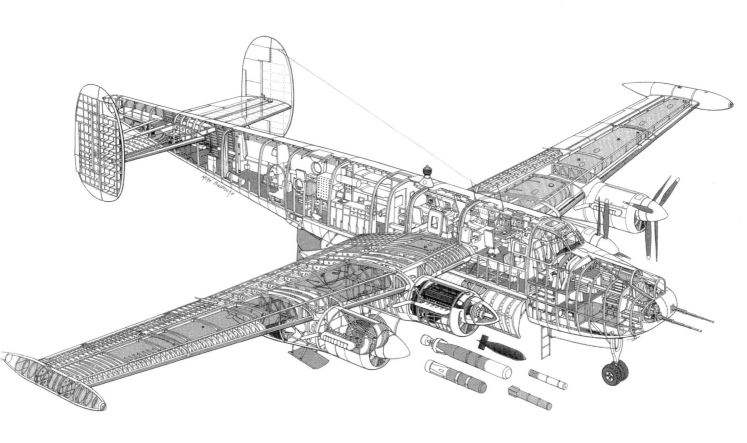

An insight into the design, construction, operation and
restoration of a classic piston-engined warbird

Keith Wilson

Contents

RIGHT **A pair of Avro Shackleton MR.2s of 42 Squadron, based at RAF St Mawgan, Cornwall, photographed in flight in 1963. The leading aircraft (right) has the ASV-13 radar stowed while the second aircraft (left) has it in the operational position.** *(Crown Copyright/Air Historical Branch image TN-4286)*

OPPOSITE **VW131 is pictured taking off at the SBAC Show Farnborough in September 1949.** *(Crown Copyright/Air Historical Branch image R-3152)*

Acknowledgements

A project of this nature requires the help and support of many people, who have contributed in different ways to make the book possible. The author would like to offer his sincere thanks to the following:

The wonderful team at the Shackleton Preservation Trust (SPT); but especially to Dave Woods, Rich Woods, Phil Woods, Richard Marriot, Barry Wheeler, Sue Guest, Mario and Michelle McLaughlin, Pete Buckingham and Mark Ward. Both the Trust and WR963 are in good, capable hands.

To the Gatwick Aviation Society and especially John Tickner for allowing me unlimited access to both Shackleton MR.3/3s – WR974/K and WR982/J – as well as providing the introduction to Mike Rankin.

To the Newark Air Museum and particularly Mike Smith, for allowing me unlimited access to Shackleton MR.3/3, WR977/B, as well as allowing me to enjoy a day exploring the excellent museum.

To the Imperial War Museum at Duxford and especially to Ester Blain, for allowing me access to photograph Shackleton MR.3/3, XF708/C, in the Preservation Hangar at Duxford.

To Alison Crook at the Museum of Science & Industry, Manchester, and Sophia Brothers at the Science Museum for providing images of Shackleton AEW.2, WR960.

To the small but enthusiastic group of volunteers at the Fenland and West Norfolk Aviation Museum (FAWNAPS) operating at West Walton Highway, near Wisbech, for access to Shackleton MR.3/3, WR971/Q.

To Darren Speechley ('Bag Man') who kindly put me in touch with a number of key Shackleton people during my research for this book.

To the various photographers who have kindly provided images to this book including: Tim Badham, Ray Barber, Andy Davey, Ray Deacon, Jonathan Falconer, Fergal Goodman, Martin Hartland, Jeremy Hughes, John James, Geoff Lee, Barrie Lewis, Peter R. March, Colin McKeeman, Brian Pace, John Parker, Martin Pole, Alexei Shevelev, Richard Vandervoord, Barry J. Wheeler, Jon Wickendon, Steve Williams, Colin Zuppich and the Avro Heritage Collection.

To Sebastian Cox at the Air Historical Branch, RAF Northolt, for providing the Branch's support and access to the collection of images; along with his encouragement and sense of humour.

My thanks must also go to Lee Barton at the Air Historical Branch for his unwavering enthusiasm, vision and attention to detail during the image selection process.

To Squadron Leader (Retired) Mike Rankin, who provided an excellent assortment of memoirs of his time on the Shackleton. Mike gave personal and first-hand accounts of learning to fly the Shackleton, as well as his experiences operating the aircraft in theatre, including time at Christmas Island in 1957 during the nuclear test programme. Mike's stories are well worth reading, so I sincerely hope he finishes the autobiography we discussed!

To Squadron Leader (Retired) John Cubberley, with whom I spent many happy and enjoyable hours learning about Shackleton operations. John had a range of first-hand information and stories to offer but, sadly, this volume is too short to include them all. Perhaps you should write your own autobiography too, John?

To Druid Petrie, a licensed engineer and former RAF technician with a wealth of experience and lots of stories to relate, complete with his own inimitable sense of humour (probably developed while spending hours on freezing-cold dispersals at various locations while servicing and maintaining Shackleton AEW.2 aircraft during his 8½ years with 8 Squadron). Druid gave of his time and support freely, and nothing seemed to be too much of a problem.

To Dave (father) and Rich (son) Woods at the Shackleton Preservation Trust; you are part of a great Team (with a capital 'T') at Coventry. Dave was always around to ensure the correct access and information was forthcoming and facilitated two visits into WR963 during engine runs. What a sound WR963 makes! To his son Rich, another hard-working and dedicated member of the SPT, who was a genuine source of information for the book. Despite never having served in the Royal Air Force, Rich has built up a wealth of Shackleton knowledge while being a major driving force in WR963's team.

At Haynes Publishing I would like to thank Jonathan Falconer, Michelle Tilling and James Robertson for their considerable input at key stages during the book's production, and for keeping me on track whenever I wavered.

Finally, sincere thanks to my wife Carol and sons Sam and Oliver. Thank you for your patience and support throughout the project. I couldn't have done it without you.

Introduction

'The Growler', 'The Old Grey Lady', 'Ten thousand rivets flying in close formation', 'The contra-rotating Nissen hut' and 'The Lovely Old Beast' are just a few of the nicknames the Avro Shackleton has been 'blessed' with over the years.

They were all terms of endearment given to her by various crew members appreciative of a truly great aircraft. The Shackleton, in her various forms, had a reputation with her crews for being oily, smelly, cold (the heating system was poor), hot (the air conditioning non-existent) and very, very noisy . . . but her crews still loved the old girl!

At this early stage, I must offer my own personal bias towards the Shackleton. I was lucky enough to have spent a most interesting week at RAF Lossiemouth in the early 1990s where I enjoyed some memorable time aboard 8 Squadron's fine Shackleton AEW.2 variant, WL757. Just a couple of years later, the Shackleton had departed RAF service, replaced by the very capable Boeing E-3D Sentry AEW.1 in service with 8 Squadron at RAF Waddington. As a Shackleton virgin, I didn't really notice the noise and the vibrations, although I do remember the friendly and efficient nature of the crew throughout the various tasks I observed while undertaking my visit.

I have recently been reminded of both the wonderful noise – the infamous Shackleton growl – of the four Rolls-Royce Griffon engines and the significant levels of vibration, when I was lucky enough to witness a number of the Shackleton Preservation Trust's regular engine runs of WR963 at their Coventry base. The first time from outside the aircraft, the second time from the inside of WR963, just behind the flight engineer's position. Even from the inside, the noise was truly amazing. Since then, on 27 September 2014, WR963 taxied – albeit briefly – around the apron at Coventry; the first step on what is hoped will be a journey to the aircraft getting air back under her wings. What a spectacular sight (and sound) that will be!

But let's stop dreaming and return to the days of RAF Coastal Command; the days of flying boats serving right across the British Empire. Prior to the Second World War, long-range maritime reconnaissance was the domain of flying

ABOVE Avro Shackleton AEW.2, WL754, of 8 Squadron photographed close to its base at RAF Lossiemouth on 16 August 1977. *(Crown Copyright/Air Historical Branch image TN-1-7736-14)*

boats – a tradition that had emerged towards the end of the First World War. In the early years of the Second World War, flying boats were at the forefront again but, as the German U-boats hunted their prey further and further beyond the effective operating radius of the contemporary flying boats – including the Sunderland – RAF Coastal Command sought to close this gap, particularly in the Atlantic waters where merchant shipping losses were especially high. The RAF solution was hastily converted Lend-Lease aircraft from the United States. Notably, they turned to the Boeing B-17 Fortress and the Consolidated B-24 Liberator.

Perhaps a little surprisingly, the land-based four-engined bombers quickly impressed the military planners. With the absence of the unavoidable weight of the flying boat's hull and with the same engine power, it provided greater payloads – especially in fuel and offensive stores. It signalled the death knell for the big military flying boat. The aircraft that particularly shone in this role was the Liberator, whose long range, combined with a reasonable weapons bay, closed the mid-Atlantic gap and provided good cover to the convoys while significantly reducing the U-boat victories.

With the end of the Second World War, Lend-Lease also came to its conclusion. Consequently, a requirement for the Liberator replacement arose. As an interim measure, Avro Lancaster aircraft coming off the production line were sent to Cunliffe-Owen's facility at Eastleigh and converted into maritime aircraft for both general reconnaissance as well as air-sea rescue (ASR) roles. Very quickly the Lancaster GR.3 became Coastal Command's standard post-war aircraft.

Unfortunately, the Lancaster was too small an aircraft to carry all of the required anti-submarine weapons and devices, and had an insufficient range for the tasks. In Bomber Command, the Lancaster had been replaced with the Avro Lincoln and it had been intended that a Mk 3 version of this would be built as the RAF's new maritime reconnaissance aircraft.

However, design studies soon found that such a conversion would not meet the requirements.

Early in 1946, the Air Staff's requirements in terms of performance, equipment and crew accommodation became apparent. Consequently, Avro decided to submit a new design to meet the Air Ministry Specification R5/46. Avro designated their new design the Avro 696 and it was not entirely devoid of Lincoln influences, having basically the same wings and undercarriage, though the fuselage was made more voluminous to house advanced Airborne Service Vessel (ASV) radar equipment and navigational aids, as well as to provide reasonable accommodation for the crew of ten, as sorties of up to 24 hours were envisaged. Three prototypes were ordered, to be equipped with Rolls-Royce Griffon 57 engines of 2,450hp each.

There was still the problem of a short-term requirement for Coastal Command until the Shackleton entered service. Once again it came from the United States, in the form of former US Navy Lockheed P2V-5 Neptunes, which would be provided to the RAF under the Mutual Defence Aid Programme (MDAP). On 27 January 1952, the first of 52 Neptune MR.1 aircraft entered RAF service with 217 Squadron, Coastal Command, at RAF St Eval, although this unit moved to RAF Kinloss in April that year. The Neptune MR.1 eventually equipped four UK-based squadrons. With a crew of ten (two pilots, two navigators, five air signallers and an engineer) and an 8,000lb weapons load, the long-range Neptunes were returned to the United States in 1957, having been replaced in service with Shackleton MR.1s.

The rest, as they say, is history. The Shackleton served on the front line with the RAF for 40 years – from April 1951 to July 1991. It served in a number of conflicts and theatres, including South Arabia and the Radfan, Suez, Borneo and the 'Beira Patrols' off the coast of East Africa. It provided important maritime reconnaissance during the nuclear test programmes around Australia and Christmas Island, assisted with trooping flights into Cyprus, provided humanitarian aid to Agadir, Morocco, after the earthquake in March 1960, relief flights to Belize after damage from Hurricane Hattie in October 1961 and to Cuba after Hurricane Flora in October 1963. The Shackleton also

ABOVE Beginning in January 1952, 52 former US Navy Lockheed P2V-5 Neptune aircraft were supplied to the RAF as Neptune MR.1 aircraft. This was a stop-gap measure until 1957, when they were replaced in Coastal Command by Shackleton MR.1 aircraft. Neptune MR.1, WX505/A-J, was photographed while on patrol in May 1953. (Crown Copyright/Air Historical Branch image PRB-1-6502)

BELOW The Fairey Gannet AEW.3 served with the Royal Navy in the airborne early warning role during the 1960s and '70s. The last of these were withdrawn from service when HMS Ark Royal docked at Devonport for the final time in November 1978. Anticipating the imminent demise of the Royal Navy's aircraft carriers, the Shackleton AEW.2 was evolved in 1971 to provide the RAF with an airborne early warning aircraft to replace the Royal Navy's AEW capability. The AEW.2 version of the Shackleton entered service with 8 Squadron in early 1972. Gannet AEW.3 – XL502/R-041 of 849 Squadron – was photographed at an airshow in 1976. (Keith Wilson)

contributed to numerous air-sea rescue operations. It soldiered on throughout the Cold War and right up to July 1991 – including a spell as an 'interim solution' to the RAF's airborne early warning (AEW) replacement for the Royal Navy's Fairey Gannet AEW.3. In all the roles it fulfilled throughout its lengthy career, the Shackleton clearly demonstrated its tough and durable nature.

The Shackleton's initial primary roles – long-range maritime reconnaissance and anti-submarine warfare – engendered an element of teamwork and camaraderie, both essential attributes for successful operations, particularly within RAF Coastal Command (or the 'kipper fleet' as they were affectionately known). It also became a troop transporter, a bomber and a search and rescue aircraft when the need arose.

Such is the affection held for the venerable old Shackleton, that she even has her own association (check out the Shackleton Association's aptly named website at www.thegrowler.org.uk). The Shackleton Association was created in 1987, just a few years before the end of the aircraft's service career. It quickly assembled a dedicated group of mainly ex-RAF members who had been associated with the Avro Shackleton during their service career and who were determined that the memory of this great aircraft should never be forgotten. They are not a preservation society as it was realised from the outset it was unlikely that they would have access to the resources to purchase and maintain an Avro Shackleton in flying order, or even to have one as a static exhibit. Nevertheless, they are keenly interested in the preservation of the aircraft for future generations.

Membership is open to anyone who is sympathetic to their aims and objectives, although the majority of members are currently ex-aircrew and groundcrew who have been associated with the Shackleton over the years. They welcome serving members of the Royal Air Force and indeed anyone who is an enthusiast of the Avro Shackleton. Through the pages of their quarterly magazine, *The Growler*, many memories are shared and numerous friends are reunited after having lost touch due to both time and distance.

A total of 185 Shackletons were constructed but, sadly, not a single example is currently flying anywhere in the world. Just a few years ago, there were two Shackleton aircraft maintained in airworthy condition. Former 8 Squadron AEW.2, WL790, departed the UK in 1994 and made its way across the Atlantic where it continued to grace the skies for a further 14 years with the Commemorative Air Force at Midland, Texas, as N790WL. Later, the aircraft underwent a complete respray into its original 8 Squadron colours and was rolled out to the public on 20 May 2013. The aircraft is now a ground exhibit at the Pima Air & Space Museum in Tucson, Arizona, and sadly is not expected to fly again.

The last 'airworthy' Shackleton MR.3 – 1722/P – is maintained by the South African Air Force Historic Flight at Ysterplaat AFB. However, although it remains 'airworthy', it has been grounded for 'safety and preservation reasons', as well as a lack of qualified crew to fly her. Apparently, there are very few hours left on the airframe's fatigue life; too few to allow a crew to be trained and kept current while still displaying it.

Enter the Shackleton Preservation Trust at Coventry Airport and their MR.2, WR963. In 1997 the Shackleton Preservation Trust was re-formed with the aim of restoring WR963 at Coventry. Within a decade of its acquisition by the new Trust, WR963 was rolling down the runway at Coventry under her own power, almost at the point of flight. Once this landmark had been reached, a feasibility study commenced to get WR963 back into the skies. WR963 has 594 flying hours remaining on the airframe, and in particular the wing spars, so there really is hope.

In 2012 the CAA was approached, and answers found to the question of the viability of flying a Shackleton in the UK. Currently, the long-standing spar life issue remains the only thing keeping WR963 on the ground. However, with flight in mind, the registration G-SKTN was reserved in February 2013 and the fund-raising began. Recently, two experienced former 8 Squadron Shackleton engineers have joined the SPT team and progress is moving up a gear. Ask any SPT team member – it is not now a question of if it flies, but when! Donations are clearly welcome to fund what will be an expensive project.

LEFT Photographed in February 1990 (during the author's visit) is 8 Squadron Shackleton AEW.2, WL757, being prepared for another sortie at RAF Lossiemouth. *(Keith Wilson)*

More information can be found on their website at www.avroshackleton.co.uk.

Much of this volume centres on the work that the SPT has undertaken on WR963 and I am particularly grateful for their considerable assistance. The sight and sound of 'the Growler' in the air will be such a welcome addition to the air display circuit, so every good wish goes to the team at SPT. I for one cannot wait for that day!

But I will leave the last word to Squadron Leader Mike Rankin RAF (Retired), who operated the Shackleton as a co-pilot, as a captain and finally a 2½-year spell as officer commanding the Air/Sea Warfare Development Unit (ASWDU) at RAF Ballykelly:

I loved the Shackleton from the first day. I had never had such a simple conversion to a new type. It fitted my 6ft 3in height very well, allowing me to stand upright in most parts of the aircraft. I was no longer strapped, immobile, to an ejector seat while encased in an uncomfortable series of life support equipments. Old-fashioned but still dramatic in appearance, even in 1958, it was better suited to the Cold War task than the Canberra light bomber I had flown on my first operational tour.

The variety of the Shackleton's additional jobs added a great deal of interest and required flexibility of mind of all crew members. In a real sense I grew up on the Shackleton as I found and coped with more and more situations never previously encountered. Within Coastal Command, training was continuous and motivation

maintained by competition – crew versus crew, squadron versus squadron, results published and winners rewarded. We could fly with a minimum crew of four on home-based pilot continuation training and when troop-carrying to Cyprus in 1958 we had six crew and 31 soldiers, a total of 37 souls on board. We dropped practice depth charges visually at 100ft above the sea and flew as high as 24,000ft in support of a scientific trial on ASWDU. We photographed electronic intelligence trawlers in the Mediterranean and within the Arctic Circle and videoed on behalf of the Atomic Weapons Research Establishment the explosions of hydrogen bombs at Christmas Island. We operated in the Atlantic and Mediterranean with other NATO forces, but also in the Far East in support of SEATO, the Southeast Asia Treaty Organization. We could also be called upon to support ground forces in protection of our bases overseas, training annually by dropping conventional 1,000lb bombs, strafing ground targets and carrying out photographic surveys.

We were kept very busy; morale in Coastal Command was outstanding and the Shackleton was where we did our stuff. It was our workplace and though we might be almost a thousand miles away on any particular day, it was also our link with home. What's not to like about the Shackleton in such favourable and interesting conditions?

Keith Wilson
Ramsey, Cambridgeshire, May 2015

Chapter One

History of the Shackleton

With a heritage stretching back to the legendary Lancaster and its successor the Lincoln, the Avro Shackleton has a distinguished ancestry. Entering service with the RAF in 1951, it operated in many roles for the next 40 years with Coastal and Strike Commands, including long-range maritime reconnaissance, anti-submarine warfare and airborne early warning. The last Shackleton AEW.2s were finally withdrawn from 8 Squadron in 1991.

OPPOSITE A beautiful study of Avro Shackleton MR.3, XF707/C, of 206 Squadron, over the British coast during a flight from RAF St Mawgan in 1964. This aircraft received the Phase 3 modifications, including the installation of a pair of Viper engines, carried out during the following year. XF707/C's service life was relatively short, however, as XF707 was withdrawn from service in April 1971. *(Crown Copyright/Air Historical Branch image T-4755)*

ABOVE Another aircraft design that provided Coastal Command with the all-important VLR anti-submarine and maritime patrol capabilities was the Boeing B-17 Fortress. In this picture, a Fortress IIA of Coastal Command's 220 Squadron was photographed 'bombing up' with depth charges at RAF Benbecula in the Outer Hebrides, just ahead of a patrol in May 1943. *(Crown Copyright/Air Historical Branch image CH-11101)*

BELOW The Short Sunderland continued to serve the RAF well after the end of the Second World War. Here, a Short Sunderland V, RN282/C of 88 Squadron, was photographed while on daylight patrol over Korea's Yellow Sea Coast in October 1950. *(Crown Copyright/Air Historical Branch image CFP-248)*

During the early part of the Second World War, RAF Coastal Command operated its Sunderland and Catalina flying boats alongside a mixed fleet of medium-range land-based aircraft such as the Hudson, Ventura and Wellington bombers.

Thanks to Lend-Lease, the RAF was able to add a formidable force of very-long-range (VLR) Boeing B-17 Fortress and Consolidated B-24 Liberator aircraft to domestically produced Handley Page Halifaxes adapted to the maritime role. Furthermore, the role of Coastal Command aircraft had changed from pure reconnaissance – chiefly the spotting of enemy surface shipping – into a combined role which now included 'hunter-killer' activities, especially against the German U-boats.

While the Sunderland and Catalina aircraft had played a vital part during the conflict – notably in

the air-sea search and rescue role – they were equipped with somewhat rudimentary 'hunting' equipment (usually the Mk 1 human eyeball), along with relatively limited armament, preventing them from effectively completing the new task.

It had been Coastal Command's Liberators in particular that had been able to spread their wings much further into the Atlantic, thereby ensuring that the German U-boats were kept below water, reducing their effectiveness against the merchant convoys. Interestingly, the maximum operational range of a fully laden Sunderland was 1,300 miles compared with 2,400 miles possible with the Liberator, while the latter was capable of carrying a significantly larger offensive payload and with a faster airspeed. Records show that over 300 German U-boats were sunk by aircraft alone, and many more by a combination of ships and aircraft.

Boeing B-17 Fortress Mks I, II and III aircraft served with 59, 206 and 220 Squadrons of Coastal Command while the Liberator equipped a total of 12 of their squadrons. Unfortunately, at the conclusion of the war, all of the aircraft provided under Lend-Lease had to be returned to the United States.

Though the hostilities were over, it was not the end of Coastal Command's responsibilities in the Atlantic. With the creation of NATO imminent, the requirement for a long-range air-sea rescue aircraft remained as great as ever. Initially, the Command turned to the Avro Lancaster to replace the US-provided VLR aircraft.

The first maritime Lancasters – designated ASR.3 (air-sea rescue) – were conversions of late production B.III bombers carried out

by Cunliffe-Owen Ltd and a few were fitted with airborne lifeboats designed by Uffa Fox. Some of these ASR.3 aircraft were later modified to the GR.3 (general reconnaissance), but later designated as the MR.3 (maritime reconnaissance) variant. The ASR and GR/MR Lancasters equipped four UK-based squadrons of Coastal Command, as well as two based at RAF Luqa, Malta. However, it quickly became apparent that there was a need for an aircraft with a greater range and payload, along with improved crew comfort. Many sorties were reaching up to and occasionally beyond 18 hours, and the limited space of the Lancaster fuselage made it difficult for the crews to work efficiently and effectively throughout these long trips. As a direct consequence of the lessons learned, Specification 42/46 was issued in March 1946, to which Avro offered the Type 696.

Avro's solution was based upon the maritime Lincoln Mk 3 and it gained an almost 'off-the-drawing-board' order on 29 March 1946 (to Specification 42/46) for an initial production run of 29 aircraft. Serials were allocated from the blocks VP254–268 and VP281–294. Three prototypes were also required and these were allocated the serials VW126, VW131 and VW135. However, a design study identified that the offered solution would not actually meet the requirements, and the team at Avro, led by chief designer Roy Chadwick, set about designing a new aircraft to which a new Specification R5/46 was actually issued on 17 March 1947.

It incorporated Lincoln-pattern mainplanes and several Tudor assemblies mated to a shorter fuselage of much larger cross-section

ABOVE After the Coastal Command B-17 Fortresses and B-24 Liberators had been returned to the US at the conclusion of Lend-Lease, the Lancaster GR.3 became the RAF's principal land-based maritime reconnaissance aircraft until the arrival of former US Navy Neptune aircraft under the MDAP Programme. Lancaster GR.3, SW324 of 210 Squadron, was photographed at RAF St Mawgan on 24 May 1952. The bomb bays of these aircraft were modified to carry an airborne lifeboat (pictured). The very last Lancaster in RAF service was a Coastal Command example, which was finally retired on 15 October 1956. *(Crown Copyright/Air Historical Branch image PRB-1-4641)*

and a high-mounted tailplane with big oval-shaped endplate twin fins and rudders. Power was supplied by two 2,450hp Rolls-Royce Griffon 57As in the inboard positions and two 57s outboard, all driving de Havilland Hydromatic six-blade contra-rotating propellers.

'Shackleton'

While much of the detailed design work on the Type 696 was carried out by others, it remained Roy Chadwick's concept. He personally supervised everything that was happening on the drawing boards, and it was he who gave the aircraft the name 'Shackleton',

BELOW An 'official' view of the prototype Avro Shackleton MR.1, VW126, taken for the Air Technical Publications Branch in January 1949. *(Crown Copyright/Air Historical Branch image ATP-17351c)*

considering it an entirely appropriate one for an aircraft which was going to voyage over vast distances. Sadly, Chadwick would not see the latest of his classic Avro 'heavies' make its first flight. He was killed on 24 August 1947 when he was on board the Tudor II prototype G-AGSU which crashed on what should have been a routine test flight at Woodford. Of the five crew on board, three were killed and one sustained severe multiple injuries; the other, the radio operator, had been positioned to the rear of the aircraft and escaped with relatively minor injuries. A later inquiry established there had been a tragic error in the rigging of the aileron controls (they had been reversed) and the aircraft crashed from 50ft into a pond on the edge of the airfield.

Those who worked on the detailed design of Type 696 recall it as one of the more straightforward tasks undertaken at the Avro Chadderton and Woodford factories; and certainly a relatively low-cost one. Don Andrew, project designer at the time, recalls: 'All we did was to slit the Lancaster and Lincoln fuselage shape from end to end and make it 2ft fatter all around.' This may have been an over-simplification, but the Lincoln wing and Tudor undercarriage components were incorporated, and the overall shape of the Manchester–Lancaster–Lincoln lineage is clear to see.

There was one fundamental change and that was the adoption of the Rolls-Royce Griffon engine with contra-rotating propellers, powerplants which had been first developed to

meet requirements for later generations of Fleet Air Arm aircraft and RAF fighters. The Griffon had already proved itself in the last marks of the Seafire, Firefly and Barracuda, and the contra-rotating arrangement went into the revolutionary Martin-Baker F18/39 fighter design which, sadly, did not see service. The Griffon was also adopted for service in some very late marks of Spitfire, including the PR.19 and F.21.

The contra-rotating propellers would help reduce crew fatigue by eliminating torque effects with applications of power, particularly during take-off, landing and manoeuvring. The engines were also fitted with a water/methanol injection system to boost the power up to 2,450hp for take-off at its (then) maximum loaded weight of 86,000lb (39,000kg).

One interesting aspect of the research and design process included sending a former RAF wartime Mitchell pilot, J.D. 'Johnny' Baker of the Avro test pilot staff, to Malta to fly with RAF Coastal Command crews flying Lancaster Mk.3s. Their views and suggestions were to be considered in respect of the new design. These views were fairly clear-cut – more room, less noise and the opportunity to consume hot food and drinks on flights lasting longer than the average civilian's two working days.

Initially the aircraft was known as the Shackleton GR.1 but it later became the MR.1. Engine installation trials were conducted at the Rolls-Royce aerodrome at Hucknall (near Derby) with two modified Avro Lancastrian C.2s (VM704 and VM728) operating with inboard Rolls-Royce Merlin engines exchanged for Griffons.

First prototype

When the first of the three prototypes (VW126) was rolled out at Woodford, the nose section included a ventral radome transparency (to house the ASV-13 scanner) under a rather blunt observation nose transparency. Located just to the rear of the nose transparency, with one on each side of the fuselage, were Boulton Paul Type I barbettes for single 20mm Hispano cannon. A Bristol Type B.17 dorsal turret was to house a pair of 20mm cannon, while a Boulton Paul Type D tail turret would contain two 0.5in Browning heavy machine guns. Under the tail was a flight

refuelling pick-up point for the 'looped line' system. The spacious bomb bay was designed to accommodate up to 20,000lb of maritime stores and ordnance. Accommodation was provided for ten crew members comprising two pilots, two navigators, an engineer and five crewmen to man the electronics gear, guns and visual lookout positions.

Adequate accommodation was to be an important aspect of the Shackleton design. Long hours spent in cramped, unsuitable conditions could seriously affect crew efficiency, especially as the time periods to be airborne were expected to be of an 'extended' nature. When designing the Lancaster, Chadwick had found the only way to provide the navigator with an adequately sized working table in the narrow, restricted fuselage was to seat him sideways to face the port wall. In the Shackleton, the same design philosophy was adopted and extended so that both navigators, as well as the sonics operator, were seated side by side at a long table. The forward-facing radar operator was positioned just aft of the sonics position so that a compact tactical team was created.

Liberal soundproofing had been incorporated following Johnny Baker's visit to Malta, while a galley was installed just aft of the main spar and forward of the mid-upper turret. The galley area also featured a rest area with bunks. To the rear of the galley was the beam position with a lookout station on either side of the fuselage. Located under the tail were cameras for reconnaissance work and weapons attack assessment. It was possible to stand upright throughout the length of the fuselage and to move freely back and forth down the starboard side behind the seats of the tactical team, although you did have to clamber over the spar positions to achieve this.

First flight

The prototype VW126 made its first flight at Woodford on 9 March 1949, with Avro chief test pilot J.H. 'Jimmy' Orrell at the controls. Initially, after the first taxi tests on the day, Orrell was not entirely happy about the amount of 'boot' required on the rudder pedals and returned VW126 for the addition of some tape-and-glue trimmers. Once completed, he

returned to the runway and after a 14-second run lifted VW126 off the ground and flew around Woodford for 33 minutes. After the engineers had completed a little more 'trimming and straking' with fabric and glue on the tail, he later made another 45-minute flight. 'Like all of Roy Chadwick's designs, it felt just right and I never had any worries,' Orrell recalled. 'I think we all knew it was going to be a good aircraft from the start. After all, it had the "Chadwick Stamp" on it.'

On 21 April 1949, VW126 was transferred to the A&AEE at Boscombe Down to begin manufacturer's trials. In an attempt to expedite

ABOVE A striking view of VW131 at the SBAC Show at Farnborough in September 1949. Note that the two port engines have been 'feathered' during this two-engined flypast. *(Crown Copyright/Air Historical Branch image R-3153)*

BELOW The prototype Shackleton MR.1, VW126, had a relatively short life, completing its flying by 1960. It then served with No 2 Radio School as 7626M for another five years. Evidence of its conversion to an MR.2 aerodynamics trials aircraft in 1951 is apparent from the nose sighting position of the aircraft. It finally met an ignominious end, having been broken up at Yatesbury in October 1965. *(Colin McKeeman)*

matters by avoiding later repetition during Ministry of Supply Service Trials, members of the A&AEE's B Squadron worked alongside the Avro test pilots.

Second prototype

The second prototype (VW131) largely resembled the first, except for the fitment of a blanking plate over the tail-end flight refuelling pick-up point. In addition, the guns were fitted into the turrets. It made its first flight on 2 September 1949 and VW131 was available to take part in the 1949 SBAC Show flying display at Farnborough in September where it was flown by Johnny Baker. On 12 February 1950, VW131 joined VW126 for the official trials preliminaries at the A&AEE. The main trials actually took place in July and the nose cannon barbettes and the tail turret were removed during this time.

Tropical trials were conducted by VW131 at Khartoum during the autumn of 1950 and on the way back the aircraft suffered a bird-strike

that smashed the radome and had a profound impact on the future design of the aircraft.

Third prototype

The third prototype (VW135) was delayed by changes required to bring the aircraft up to as near production standards as possible. It eventually flew for the first time on 29 March 1950, one day after the first production aircraft (VP254) took to the air! VW135 was devoid of nose barbettes, flight refuelling pick-up point and the tail turret. In addition, the window at the navigation position just ahead of the port wing had been deleted. The aircraft joined the official trials at Boscombe Down in July 1950 and continued the trials until March 1954, being used mainly for armament development.

Into production

The urgency of getting some Shackletons into squadron service necessitated the immediate construction of production aircraft. As stated earlier, the first production aircraft (VP254) flew on 28 March 1950 and was quickly followed by VP255, which flew on 30 June, in plenty of time to attend the RAF Display at Farnborough on 7 and 8 July as a static exhibit. In November it went to Boscombe Down for trials on the Bristol Type B.17 turret.

Next to fly was VP257 on 28 August and it carried out displays at the 1950 SBAC Show at Farnborough in the hands of Johnny Baker. On 30 December it went to the Central Servicing Development Establishment at RAF Wittering for four months.

VP256 flew on 18 September and was the first Shackleton delivered to the RAF, being allocated to RAF Manby on 28 September for

the preparation of the Pilot's Notes. The next two aircraft, VP258 and VP259, flew on 13 and 24 October respectively.

Deliveries commence

All 29 of the first order under contract 6062-4 were delivered during 1950–51. The first unit to receive aircraft, starting in March 1961, was 120 Squadron, temporarily based at RAF Kinloss so that service trials could be conducted in liaison with 236 OCU, which was in the process of forming there. Three early production aircraft also went to the Air/Sea Warfare Development Unit (ASWDU) at RAF St Mawgan so that weapons developments and installation trials could start.

In October 1951, a slight redesign was made with the outer nacelles being widened to enable Griffon 57A engines to be fitted all round for standardisation. A further 58 aircraft were ordered under Contracts 3628 (WB818–861) and 5047 (WG507–558). The modified aircraft were designated Shackleton MR.1A and all earlier MR.1s were later brought up to this standard during 1955–56. All MR.1A aircraft were delivered in 1952 and further squadrons were equipped as new aircraft became available.

It was soon discovered that the 'ample soundproofing' that had been such a prominent feature of the prototypes was omitted

RIGHT Shackleton MR.1A, WB822, was retained by the manufacturer to allow it to be demonstrated at the SBAC Show at Farnborough in September 1951 where it put on a great display during the daily flying programme. It was not delivered to its first unit – 236 OCU at RAF Kinloss – until 28 December 1951. *(Crown Copyright/Air Historical Branch image PRB-1-3375)*

ABOVE Early deliveries of Shackleton MR.1s were made to 120 Squadron at RAF Kinloss in March 1951, then to the ASWDU at St Mawgan and later to 236 OCU at Kinloss. An unidentified MR.1 coded 'B' (thought to be VP284) was photographed while operating with 236 OCU at RAF Kinloss. *(Crown Copyright/Air Historical Branch image PRB-1-9531)*

BELOW The Sara airborne lifeboat was fitted to prototype Shackletons and tested. The arrangement was discarded at an early stage in favour of the much more efficient Lindholme Gear. Here, lifeboat No 801 is seen fitted to an early MR.1. *(Crown Copyright/Air Historical Branch image PRB-1-10733)*

completely from production aircraft. Although the crews enthusiastically welcomed the roomy interior, large weapons bay and ample power from the four Griffon engines, the interior was *very* noisy, and together with the matt-black interior paintwork, created a difficult and depressing environment in which to work.

Official trials had shown that the radar scanner efficiency was below that expected and – coupled with its vulnerability to bird-strikes in the chin position – it was decided to reposition the scanner. At the same time the opportunity was taken to streamline the nose shape and incorporate a turret. Although submarine deck guns were a thing of the past – following their quest for more underwater speed – and 'scare guns' had been made redundant, it had already been anticipated that the Shackleton would be deployed in a secondary role of colonial policing and therefore a turret would be of significant benefit.

Enter the Shackleton MR.2

Issue 2 of Specification R5/46, dated 3 July 1950, gave the go-ahead and the first prototype (VW126) was rebuilt during

the winter of 1950/51 to incorporate the new ideas. It emerged in the summer of 1951 as the aerodynamic prototype of the MR.2 aircraft. The nose had been lengthened and now incorporated a dummy turret and lookout position, while the rear fuselage had been lengthened to finish in a cone shape. Twin retractable tailwheels had been fitted, and lockable rudders and toe brakes had been added to provide better control on the ground. A dummy radome had been positioned ventrally aft of the wing.

The first flight in this form was made on 19 July 1951 and the aircraft went to Boscombe Down for trials. One of the MR.1A production aircraft was taken from the production line and rebuilt to full MR.2 specification as the prototype aircraft. The nose incorporated a Boulton Paul Type N turret with two 20mm Hispano cannon and, under it, a prone bomb-aimer's position with an optically flat screen. The new tail cone was also transparent, to provide a prone lookout position for weapons attack assessment. The radome was now a semi-retractable 'dustbin-like' affair.

WB833 made its first flight on 17 June 1952. The following month it went to the

RIGHT The first 'full prototype' of the Shackleton MR.2, WB833, photographed in flight during October 1952. The MR.2 featured a redesigned and lengthened nose and a new elongated tail with observation post to aid visual searches. The underbelly 'dustbin' containing the ASV-13 search radar is fully extended in this image, but could easily be retracted back into the fuselage and stored. *(Crown Copyright/Air Historical Branch image PRB-1-5544a)*

A&AEE at Boscombe Down for three months of engineering, handling and radio trials.

The trials proved successful, and the last ten aircraft of the MR.1A order were built to MR.2 configuration (WG530–533 and WG553–558). This restricted the total of MR.1 and MR.1A aircraft to just 76. However, further orders for MR.2 aircraft were placed in 1952 under Contract 6129 (WL737–801) and later under Contract 6408 (WR951–990).

The first three MR.2 aircraft were completed in September 1952 and were loaned for trials work to the Ministry of Aviation (MoA), as the Ministry of Supply (MoS) had now become known. WG530 was delivered to the A&AEE at Boscombe Down on 25 September 1952. WG533 was the first MR.2 delivered to the RAF, going to RAF Manby in October for the completion of the Pilot's Notes, while WG532 went to the ASWDU in January 1953 where it was fitted out for rocket-firing trials. Four rocket rails were installed under each outboard mainplane but the trials were not a success as the aircraft proved too heavy on the controls to manoeuvre for rocket attacks. Consequently the idea was dropped.

MR.2 development work

Development of equipment for the Shackleton continued with the ASWDU, and an MR.2 (WL789) was fitted with an extended tail for Magnetic Anomaly Detection (MAD) trials. Unfortunately, the airframe proved to be electronically unsuitable for the equipment, and the heaviness of the controls made effective tactical use of the equipment unworkable. Although the trials were extended to include overseas operations in Malta and WL789 retained the equipment, the project was shelved.

Another development by the ASWDU concerned illumination for night attacks.

ABOVE After completing manufacturer's trials followed by A&AEE performance trials, Shackleton MR.2, WB833, was delivered to the ASWDU in November 1960 where it operated as 'B'. The aircraft was eventually lost when it crashed off the Mull of Kintyre in April 1968. *(Colin McKeeman)*

BELOW Shackleton MR.2, WL789, made its first flight on 10 June 1953 and was modified to carry a MAD (Magnetic Anomaly Detection) boom in a stinger tail. It was delivered to the ASWDU in September of the same year for trials, which were completed by April 1958. The modified tail was then removed and the aircraft delivered to 224 Squadron. The aircraft is seen here with the modified tail and painted as 'F-D' while serving with the ASWDU. *(Colin McKeeman)*

BOTTOM Shackleton MR.2, WL796, made its first flight on 23 August 1953. It was exhibited at the SBAC Show at Farnborough later in the year equipped with an airborne lifeboat for air-sea rescue, but the aircraft never deployed it operationally. WL796 was delivered to 38 Squadron in January 1954. *(Crown Copyright/Air Historical Branch image AHB-MIS-AVRO-480-017)*

The standard illumination equipment of the Shackleton was four six-barrelled dischargers which fired flares from the starboard beam, each flare giving 3 seconds' illumination at 3.5 million candle power. In an attempt to improve illumination, trials were conducted using rocket-fired flares called 'Glow Worm', the idea being to place the illuminants behind the target for silhouetting. The flying techniques required for proper positioning of the flares proved difficult and hazardous – even when attempted in the relatively nimble Neptune trial aircraft. Consequently, all plans for using the idea in the proposed MR.3 aircraft were dropped, although they were all eventually delivered with full rocket circuitry in the wings.

The Shackleton had shown from the very start, by results in exercises with the Royal Navy and with NATO forces around Europe, that it was an excellent maritime reconnaissance platform. However, working conditions for the crew were far from ideal and complaints received from the squadrons eventually led to fatigue trials being conducted at the Royal Aircraft Establishment (RAE) at Farnborough in 1953 and 1954. As more Shackleton aircraft were soon to be required to replace the ageing fleet of Sunderland flying boats, construction of the MR.2 was terminated at WR969 in September 1954. A complete redesign of the aircraft was undertaken, improving and modernising it as the Avro Type 719 – the Shackleton MR.3.

Shackleton MR.3

Issue 3 of Specification R5/46 covering the new type had been released on 18 November 1953, but work did not start until 1954. The new design featured a tricycle undercarriage with twin wheels and hydraulic brakes, which required a minor change to the nose contouring and a sharpening of the weapons bay to fit the retractable nosewheel and crew entrance hatch. The wing plan form was changed and modified ailerons of greater chord were fitted to improve handling. Permanently fixed wingtip tanks were added to provide an increase in fuel capacity to 4,248 imperial gallons, and a fuel jettison system was added.

The dorsal turret was removed, enabling the crew rest room to include full cooking facilities and be incorporated aft of the new wing spar. The room was sealed fore and aft by bulkheads with access doors on their starboard side. A clear-vision Perspex cockpit canopy was fitted to improve pilot vision and all equipment 'black boxes' were installed into racks at roof level, behind drop-down doors. The entire aircraft was liberally soundproofed and lined in brown and cream Rexine.

To increase the efficiency of the tactical team, the table down the port wall was extended aft to the rest room bulkhead, to include the radar operator. Large, comfortable armchairs were fitted at all crew stations.

To improve crew comfort still further, the old

noisy engine exhaust stubs were replaced by a sealed unit which deflected the gases (and noise) under the wing. There was very little new equipment actually fitted in the aircraft, but account had been taken of new equipment in the design stage, and aerials and wiring had been incorporated for later installations.

Airborne lifeboats had been developed for search and rescue (SAR) use and had been a feature of the Lancaster ASR.3. Similarly, all Shackletons were equipped to carry them. However, the Lindholme Gear – a set of five rope-connected cylindrical containers with a dinghy in the middle one – was preferred. Consequently, the lifeboat was never used operationally by the Shackleton in RAF service.

The prototype Shackleton MR.3 (WR970) made its first flight on 2 September 1955, and appeared at the SBAC Show at Farnborough in the same month. Unfortunately, the aircraft suffered from unsatisfactory stall characteristics, especially with the bomb doors open, and as a consequence underwent prolonged handlings trials with the A&AEE at Boscombe Down. WR970 was returned to Avro in November 1956 for a modified stall warning device to be fitted before further trials could be undertaken.

Prototype MR.3 lost

On 7 December 1956, during further stall tests, control was lost during an induced stall and the aircraft became inverted in cloud. Although control was eventually regained, there was no power available as the engines had oiled up and the windmilling propellers caused the aircraft to stall into the ground near the village of Foolow in Derbyshire. The aircraft was destroyed in the subsequent crash and all four crew members died.

The second MR.3 (WR971) did not fly until December 1956 and, as a consequence, entry of the variant into service was delayed. The aircraft was eventually fitted with 'vee' additions to the leading edge of the inboard section of the mainplane, which enabled that part of the wing to stall first, thereby providing a wings-level stall.

Deliveries to RAF squadrons began in August 1957, with 220 Squadron at RAF St Eval being the first to receive the type. The squadron moved to St Mawgan a month later as the longer runway there was more suitable for the new aircraft and service trials commenced. No 206 Squadron moved to St Mawgan on 14 January 1958, partially re-equipped with the MR.3 aircraft due to the slow delivery rate and joined in the service trials. Following delays caused by problems with air entering the fuel system and hydraulic failures related to piping, the trials were eventually concluded and further squadrons started re-equipping with the new aircraft.

Shackleton trainers

In 1956 it was decided to standardise on the Shackleton aircraft for RAF Coastal Command and plans were made to phase out the remaining Lockheed Neptune aircraft. At the same time, the opportunity was taken to reorganise the training of aircraft crews. Up to this time, aircrew had spent ten weeks at the School of Maritime Reconnaissance at St Mawgan flying Lancaster Mk.3s, and then proceeded to 236 OCU at RAF Kinloss to convert on to their assigned aircraft.

ABOVE One of the 'official' images of the Shackleton MR.3 prototype, WR970, taken at Avro's Woodford facility in April 1956. The aircraft features the long tube fitted to the nose carrying instrumentation for the yaw meter. Sadly, the aircraft was destroyed in a fatal accident in Derbyshire on 7 December 1956, killing all four on board including Squadron Leader Jack Wales, a senior Avro production test pilot. *(Crown Copyright/Air Historical Branch image ATP-28858b)*

ABOVE Once the MR.3 aircraft had entered service alongside the earlier MR.1s and MR.2s, the need became apparent for a dual-control training variant for service with the successor of 236 OCU, the Maritime Operational Training Unit (MOTU). This need was met by modifying some of the early MR.1s and later MR.1As. WB831 had originally been built as an MR.1A in 1951 but was converted to T.4 configuration in August 1956 and delivered to MOTU in April 1958 where it served as 'S'. *(Colin McKeeman)*

BELOW Another MR.1A converted to T.4 configuration was WB844. After conversion in July 1956, it was delivered to MOTU in January 1958 where it served as 'R'. *(Colin McKeeman)*

The whole training process was amalgamated under a single roof at Kinloss called the Maritime Operational Training Unit (MOTU) and special training versions of the MR.1A were ordered. Designated Shackleton T.4, these accommodated additional radar positions in the bunk area. This was possible as the mid-upper turrets had been removed from the MR.1/MR.1A and MR.2 aircraft during 1955–56 to provide more internal space.

The first two Shackleton T.4 aircraft (WB837 and WG511) entered service with MOTU in August 1957. Eventually 17 such conversions were completed.

During another later reorganisation, MOTU moved down to St Mawgan in July 1965 and the long-serving T.4 aircraft were eventually replaced during 1968 with training versions of the MR.2 Phase 3 aircraft – designated T.2. Ten such aircraft were converted by Hawker Siddeley at Langar during 1967.

MR.3 upgrades

Maritime reconnaissance work is primarily centred on the task of combating the submarine menace, so the aircraft equipment must be continuously upgraded and renewed to achieve this. Initially, Shackleton crews had to rely on the ASV-13 radar scanner and eyesight for detection, non-directional sonobuoys for investigation and depth charges for attack. Over the years, equipment was added for detecting and homing on to engine exhausts (via a scoop of the port nose side), as well as directional sonobuoys for better underwater tracking and

LEFT VP293 made its first flight on 18 July 1951 and entered squadron service. In August 1956 it was converted to T.4 configuration and delivered to the MOTU. In January 1964 it was acquired by the Royal Aircraft Establishment (RAE) and continued with trials work until withdrawn from service in May 1975. It was subsequently sold to the Strathallan Museum and flown to Scotland in May 1976 where it remained on display until broken up on site following the failure of the museum. It was photographed at Strathallan still wearing its RAE colours prior to the museum's demise. *(Martin Stephen)*

homing torpedoes for attacking submarines at greater depths. The MR.3 had introduced a sonics plotting table and around the late 1950s much new equipment was becoming available. Consequently, a modification programme was commenced in 1959 to completely upgrade the aircraft. For ease of conversion, it was decided to complete this process in three distinct phases that eventually took until 1968 to accomplish.

The squadrons were kept at full strength during the modification programme by swapping aircraft as modified airframes became available from the maintenance units and aircraft changed squadrons on several occasions during this period. To assist the programme, 203 Squadron converted on to MR.2 aircraft in 1962, to provide a pool of MR.3s available for conversion.

Phase 1 introduced the long-awaited ASV-21 radar, ILS (instrument landing system), VHF radio homers, radio/radar altimeters and Doppler navigation equipment, along with search and rescue homing equipment. The Mk 10 autopilot was fitted (the MR.3 had been hand-flown previously) along with an improved intercom, a flame-float dispenser in the beam position and the wiring for the lifeboat was removed. The first Phase 1 aircraft were delivered to squadrons during the summer of 1959. The MR.2 aircraft also received the sonics plotting table from the MR.3 aircraft and for a while these were known as MR.2C aircraft on the squadrons.

Phase 2 modifications changed the outline of the upper fuselage by introducing electronic counter measures (ECM) equipment, the aerial being the large, white 'spark plug' on the fuselage top. Other equipment included UHF radios, UHF radio homer, TACAN (radio bearing and distance equipment), active sonobuoys and improved radio compasses. The first Phase 2 aircraft arrived on the squadrons in late 1961. The MR.2 aircraft also received the MR.3 engine exhaust system.

Phase 3 brings jet power

The Phase 3 modifications involved substantial structural rebuilding and changed the outline of the MR.3 aircraft. The significant amount of added equipment had eroded the performance of the MR.3 aircraft so much so that additional power was badly needed at take-off. Bristol Siddeley Viper jet engines were fitted into the outer engine nacelles, below and behind the Griffons. This required strengthening the wing spars, and at the same time the wing fuel tanks were enlarged slightly and the wings were re-skinned with the (unused!) rocket wiring being removed during the process.

Internally, the tactical table was extended aft to include another position, which took up part of the crew rest room. This was remodelled to have a rear-facing dinette and just one tier of three bunks. All aircraft received the improved radio compass and VHF radio. The navigation equipment was significantly improved and the new GM.7 was fitted for enhanced accuracy. All this new electrical equipment required extra electrical power and a crate of seven inverters was therefore fitted in the nose.

The empty weight of the MR.3 had risen by 6,500lb and the maximum take-off weight by 8,000lb from the original MR.3 aircraft.

The first MR.3 Phase 3 aircraft was WR973 and it flew for the first time on 29 January 1965. However, the aircraft was found to be

ABOVE Shackleton MR.3, WR972, made its first flight on 6 November 1961. Its first task was trials work with the A&AEE at Boscombe Down, including work on the Autolycus system. It briefly returned to Woodford before being officially acquired by the MoA for use by the RAE. It spent the remainder of its life engaged on a variety of trials work including sonobuoy, mixed bomb load clearance and the Orange Harvest ECM. In this photograph, taken on 14 September 1968, it was a welcome visitor to the Battle of Britain Day at RAF Coltishall. *(Steve Williams)*

tail-heavy and all the soundproofing aft of the crew rest area was removed, the inside instead being sprayed with green flock! Phase 3 aircraft started to arrive on squadrons in the spring of 1965 but these were without the additional Viper engines, which were retrofitted during 1966–67.

Perhaps a little surprisingly, the Viper jet engines ran on the normal gasoline (Avgas) fuel system rather than an appropriate jet fuel. As a consequence they had a very short lifespan, being used only for take-offs and overshoots. Recorders were fitted to monitor their running time and eventually a Time Before Overhaul (TBO) of 250 hours was reached, with an operational necessity maximum of 275 hours. All operating controls for the Vipers were fitted on the engineer's desk under a hinged transparent Perspex lid.

MR.2 upgrades

During the Phase 3 upgrades, the remaining MR.2 aircraft were completely rebuilt inside up to full MR.3 standards. As they did not have the extra weight of the tricycle undercarriage, though, they did not require the Viper installation, and therefore were able to retain their fuselage soundproofing.

The MR.2 Phase 3 aircraft began arriving on the squadrons during the autumn of 1966. All Phase 3 aircraft had the ability to drop nuclear depth charges while the Griffon engines had become the Mk 58 as a result of a slight change to the oil feed.

BELOW Avro's only export success with the Shackleton was an order for eight MR.3 aircraft in March 1954 from the South African Air Force, replacing their Sunderland GR.5s in service. Delivery of the first two aircraft was made in May 1957, while all eight had been received by the SAAF's 35 Squadron at Ysterplaat, near Cape Town, by February 1973. All had been withdrawn from service by November 1984 although a good number were successfully preserved. This example, 1722, was retained in an airworthy condition at the SAAF Museum for a number of years. *(Peter R. March)*

Export order from South Africa

In the early 1950s the South African Air Force began to seek a replacement for its fleet of Sunderland aircraft which maintained a watch over the Cape of Good Hope's shipping lanes. The Sunderland aircraft served faithfully until 1957 but before then the South African government had been impressed by two goodwill visits of RAF Shackleton squadrons and became interested in the development plans for the MR.3 version.

An order was placed for eight MR.3 aircraft, initially in Phase 1 form (later to be upgraded to Phase 2 and eventually to Phase 3), but for a number of reasons the additional Viper engines were never fitted. One of the reasons for the absence of the Vipers was that the 6,000ft runway requirement of the Royal Air Force did not apply in South Africa.

The first two Shackleton MR.3 aircraft (1716 and 1717) were handed over to the SAAF on 21 May 1957, after a party from 35 Squadron, SAAF, had undergone an intensive familiarisation course on the aircraft at Woodford.

Eventually all eight aircraft (1716–1723) were delivered and they actually entered service 12 days ahead of the MR.3 in RAF service! They also continued in service until November 1984 – much longer than the RAF aircraft. Perhaps the decision not to add the Viper jets made a big difference to fatigue; clearly less strain was imposed on the airframes without the additional engines.

Stop-gap AEW

As the 1960s drew to a close, several projects were advanced in an attempt to solve the airborne early warning problem facing Britain. The Royal Navy had a small fleet of Fairey Gannet AEW.3 aircraft, but these were carrier-based and a decision to phase out conventional fixed-wing aircraft carriers had been taken by the government of the day as a cost-saving measure. The last carrier, HMS *Ark Royal,* would eventually dock at Devonport for the final time in 1978.

The RAF looked to find a short-term solution to the problem and, once again, the venerable

Shackleton offered an interim answer. The plan was to convert 12 surplus MR.2 Phase 3 Shackleton aircraft into AEW.2 variants, the conversion work being undertaken by Hawker Siddeley at Woodford and Bitteswell. At the time, it was considered a strictly short-term arrangement as the new Nimrod AEW.3 was scheduled to become available in the early 1980s; it certainly was not expected that the aircraft would have to continue in the role right up to July 1991, when the Sentry AEW.1 eventually joined the RAF.

The Shackleton MR.3 fleet was considered unsuitable for the role, with its additional operating weights and Viper engines leaving little or no fatigue life on the airframes, all of which had been retired by the end of 1971.

The remaining Shackleton MR.2 fleet was scoured and those airframes with the longest remaining fatigue life were selected for the conversion programme and delivered to Hawker Siddeley. At the time, the choice of airframes was good as Nimrod aircraft were replacing the Shackletons in maritime reconnaissance squadrons. At the same time, surplus AN/APS 20 radar systems from the Fairey Gannets being withdrawn from Royal Navy service were becoming available.

NATO had decided to purchase the Boeing E-3 Sentry, but the British government decided to go it alone with the development of their own long-term AEW solution (the Nimrod AEW.3 was already in development), so the Shackleton AEW.2 would merely be expected to provide an interim solution.

In simplistic form, the AEW.2 conversion involved removing the ASV-21 radar from the rear of the fuselage and fitting second-hand AN/APS 20 radar systems under the nose of the aircraft at the forward end of the bomb bay. (A more detailed examination of the project will be covered in Chapter 4.)

The first conversion (WL745) underwent its maiden flight as an AEW.2 on 30 September 1971. Then, on 1 January 1972, 8 Squadron was re-formed at RAF Lossiemouth to become the first AEW squadron in the RAF, although it initially operated at nearby Kinloss as their home runway was undergoing repair. Many of the radar operators on 8 Squadron were former Royal Navy AEW aircrew. The first Shackleton AEW.2 to join the unit was WL747, which arrived on 11 April 1972 with eight aircraft on strength by the end of the year. The squadron moved from Kinloss to Lossiemouth once runway work had been completed, and eventually the squadron had 12 AEW.2 conversions on strength, along with a selection of MR.2 and T.2 aircraft for crew training.

In the 1981 Defence Review, 8 Squadron's allocation of Shackleton aircraft was reduced to just six, which left them particularly short. It had been expected that deliveries of the Nimrod AEW.3 would commence in 1982, but the programme was fraught with difficulties and was eventually cancelled in 1987, allowing an order to be placed with Boeing for seven Sentry AEW.1 airframes. These finally began arriving at RAF Waddington in November 1990 and the following year the final Shackleton AEW.2 sortie took place – bringing to an end a 19-year 'interim solution'.

Chapter Two

Shackleton at peace

──●──

The main task of Coastal Command Shackletons was maritime reconnaissance (MR) and anti-submarine warfare (ASW). Search equipment for the role was the ASV-13 radar, which was capable of picking up a decent target at a range of up to 40 miles away, assuming an altitude of around 1,000ft along with favourable sea conditions.

OPPOSITE Avro Shackleton MR.2, WR960/X, of 228 Squadron, RAF Coastal Command, in flight near its base at RAF St Eval. WR960 was converted to AEW.2 configuration in 1971 and operated with 8 Squadron at RAF Lossiemouth. It was withdrawn from service in November 1982 and is now exhibited in the Museum of Science & Industry in Manchester. *(Crown Copyright/Air Historical Branch image T-326)*

ABOVE Shackleton MR.1, VP256. This aircraft was used by the RAF Handling Squadron at RAF Manby in 1950–51 while the Pilot's Notes were written. Following completion of the work, VP256 was delivered to 224 Squadron in August 1951. The aircraft was involved in a take-off incident on 26 October 1954 at RAF Ballykelly when the pilot attempted to take off with the elevator locks engaged. The aircraft was struck off charge and dumped at 23 MU, RAF Aldergrove, before being scrapped in February 1963. *(Crown Copyright/Air Historical Branch image PRB-1-6180)*

If required, an attack could be made with depth charges or, later, acoustic torpedoes. The Shackleton also carried a large variety of pyrotechnics – flares and marine markers – in the bomb bay. Also carried in the rear of the bomb bay was the Lindholme Gear, particularly useful when the aircraft was operated in the SAR role.

It is difficult to describe a 'typical' sortie but, for the Coastal Command squadrons based at RAF Ballykelly in the days of the MR.1 and MR.1A aircraft, it would probably involve a Navigation Exercise (NAVEX) of up to 15 hours over a triangular course across the North Sea, with operations usually down at 1,000ft or lower, in all weathers, day and night. Most sorties finished with a practice homing and simulated attack on a radar buoy moored off the northern coast of Ireland. This would be carried out at 300ft at night and down to just 100ft in daylight.

As more aircraft became available to the various squadrons, bombing and gunnery practice was added to the rota, often while aircraft were detached to other bases. SAR was carried out in rotation by individual crews, and was a very important task.

Conditions on board the Shackleton were noisy and uncomfortable. The long flights over the water were often tedious, but the Shackleton was a sturdy, capable aircraft and proved itself to be an excellent hunter of submarines.

Into service

No 120 Squadron at RAF Kinloss was the first unit to be equipped with the Shackleton, receiving airframes from the initial

RIGHT No 120 Squadron at RAF Kinloss was the first unit to equip with the Shackleton, receiving airframes from the initial production batch of MR.1s in April 1951. The first aircraft to arrive was VP258, which touched down at Kinloss on 3 April. It was quickly followed by four more – VP259–262. Seen here is MR.1A, WB828/C, which joined 120 Squadron in September 1955. *(Colin McKeeman)*

Poor sea conditions could, however, severely restrict the effectiveness of the radar return. Once confirmation of an underwater contact was established, a pattern of sonobuoys would be laid over the location and the actual position of the underwater target could be deduced from the sounds picked up by the sonobuoys. At this stage in the Shackleton's early life, sonobuoys were only of the 'passive' variety; they only received sounds from other sources (i.e. the target) and did not transmit signals that could be bounced back from an underwater object.

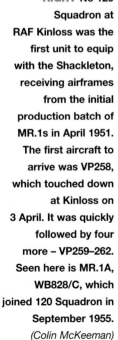

production batch of MR.1s in April 1951. The first aircraft to arrive was VP258, which touched down at Kinloss on 3 April. It was quickly followed by four more: VP259–262. Also based at Kinloss was 236 OCU, which formed on 31 July 1957. VP264 was the first Shackleton MR.1 delivered to the OCU on 31 May 1957. By the end of the year they had a complement of 12 aircraft. Service trials of the Shackleton were conducted by both units operating together. It was an excellent arrangement having both the OCU and first operational squadron on the same base as it permitted the natural feedback into the OCU of ideas and techniques developed by 120 Squadron.

Subsequent deliveries were made to the ASWDU at RAF Ballykelly on 27 April 1951, although this unit moved to RAF St Mawgan the following month. The next squadron to receive MR.1 aircraft was 224 at RAF Gibraltar, which accepted its first aircraft in July 1951 to replace the ageing Halifax GR.6, the conversion being completed by October.

Next in line was 220 Squadron, which had re-formed at RAF Kinloss on 24 September 1951 and received its first aircraft – VP294 – on the same day. On 1 January 1952, 269 Squadron re-formed at RAF Gibraltar from the nucleus of 224 Squadron. They took delivery of their first aircraft shortly afterwards but moved to RAF Ballykelly in March. Also resident at Ballykelly was the Joint Anti-Submarine School (JASS), being delivered its first aircraft, an MR.1A, WB849, on 18 March 1952.

After completing service trials, 120 Squadron moved from Kinloss to its permanent base at RAF Aldergrove in April 1952. On 1 May that

ABOVE Shackleton MR.1A, WB831, was issued to 220 Squadron at RAF St Mawgan on 21 December 1951 where it was photographed on 24 May of the following year. Shortly afterwards, the aircraft was coded T-Q. In 1956, WB831 was converted to a T.4 training aircraft with additional internal radar positions for instructors and students and delivered to the MOTU in November 1957. *(Crown Copyright/Air Historical Branch image PRB-1-4640)*

year, the re-formed 240 Squadron joined them there, taking VP255 on charge the same day before moving to Ballykelly the following month. Meanwhile, 42 Squadron was re-formed at RAF St Eval on 28 June 1952, their first aircraft – WG509 – having been delivered back in March.

BELOW An unidentified Shackleton MR.2 of 224 Squadron photographed while flying over its base at Gibraltar. Shackleton MR.2 aircraft began deliveries to 224 Squadron in May 1953 and remained with the squadron at Gibraltar until it was disbanded in October 1966. Somewhat unusually, this aircraft carries the code '3' on the nose. *(Crown Copyright/Air Historical Branch image CMP-898)*

RIGHT In November 1960, to advertise a forthcoming airshow at Khormaksar, Aden, the station parked a selection of its resident aircraft in the shape of the letter 'K'. The station, which was opening its gates to the public for the first time since the arrival of the RAF 35 years earlier, was home to squadrons of Beverley, Shackleton, Valetta, Twin Pioneer, Hunter, Meteor and Canberra aircraft, along with Sycamore helicopters. *(Crown Copyright/Air Historical Branch image CMP-1151)*

BELOW A different view of two versions of the Shackleton operating with squadrons based at RAF St Eval, 14 May 1954. MR.1A, VP293/A-F of 42 Squadron is nearest the camera, and an unidentified MR.2 (coded B-Z$_1$ and thought to be WL742) of 206 Squadron, fly in loose formation over a Royal Navy vessel. *(Crown Copyright/Air Historical Branch image X-50909)*

BELOW On 30 August 1955, RAF Thorney Island hosted a series of helicopter and search and rescue media demonstrations including this 'rescue' scenario involving a small group of 'ditched sailors' along with a 42 Squadron Shackleton MR.2, WL745/A-E. *(Crown Copyright/Air Historical Branch image PRB-1-10328)*

LEFT On 31 May 1956, 18 Shackletons from 42, 120, 204, 220 and 228 Squadrons formed the Queen's birthday flypast flying over Buckingham Palace at 1,000ft in three separate formations. This photograph shows the last aircraft passing over Buckingham Palace during a very bumpy flight that had been hard work for all the crews involved. *(Crown Copyright/Air Historical Branch image PRB-1-11858)*

Finally, 206 Squadron was re-formed at St Eval on 27 September 1952 and received the first MR.1A, WB833, on 11 October.

At this stage, seven Shackleton squadrons existed and these were being supplemented in Coastal Command with Neptune squadrons that were becoming operational at the time. The squadrons were soon busy with their new aircraft – gaining time and experience on them – and it wasn't too long before they were called in to active service.

Eventually, the Shackleton in its various guises served with 16 operational squadrons and four training flights in the RAF, as well as with the South African Air Force.

ABOVE RIGHT No 206 Squadron Shackleton MR.3s at RAF St Mawgan in December 1959. *(Crown Copyright/Air Historical Branch image PRB-1-18247)*

RIGHT A nice study of Shackleton MR.3, XF707/P of 201 Squadron, based at RAF St Mawgan, photographed in November 1960. *(Crown Copyright/Air Historical Branch image PRB-1-20153)*

ABOVE Shortly after entering service, the Shackleton was soon involved in a number of NATO detachments, as well as exercises and goodwill tours. The aircraft was well suited to this kind of work as it was able to carry its own groundcrew and a good supply of spares in both the large fuselage as well as within luggage panniers located in the bomb bay. The first tour was undertaken by Shackleton MR.1As of 220 Squadron who visited Ceylon between February and April 1952. Five of the squadron's aircraft were photographed close to Negombo, Ceylon, on 24 April 1952. These are WB821/T-L, WB823/T-N, WB825/T-M, WB831/T-Q and WB837/T-B1, although this code was later changed to 'T-S'. *(Crown Copyright/Air Historical Branch image CFP-559)*

BELOW Four Shackleton MR.2s of 42 Squadron also visited Ceylon in 1953. This is WG556/A-J. *(Crown Copyright/ Air Historical Branch image CFP-737)*

Goodwill ambassadors

The Shackleton was a rugged aircraft able to support itself while away from base. The large fuselage made it possible to carry groundcrew and spares, while tools and equipment could be loaded on to panniers and carried in the large bomb bay. As a result, the Shackleton was chosen to undertake numerous detachments to NATO member countries for exercises, along with a variety of goodwill tours.

Part of the British commitment in the Far East was an annual Fleetex (Fleet Exercise), in which the Indian and Pakistani naval forces co-operated with the Royal Navy (the RAF providing maritime support aircraft from Coastal Command or the Middle East Air Force). In previous years this had involved Lancaster aircraft, but now it was the turn of the Shackleton, and 220 Squadron spent January 1952 preparing for the aircraft's first visit to the area.

Six aircraft left St Eval on 8 February and routed via El Adem and Khormaksar, arriving at Negombo in Ceylon for Operation Scull on 13 February. The following day they commenced exercise flying, which consisted mainly of anti-submarine patrols around the fleet, but the Shackleton also shadowed HMS *Gambia*, which was operating as an 'enemy' cruiser, and completed the detachment by flying five sorties on low-level strikes against the combined East Indian Fleet.

This was followed by a detachment of 224 Squadron for another Fleetex visit to Ceylon in August of the same year.

Yet another Fleetex visit was planned to Ceylon for April 1953 and this time 42 Squadron was selected. No 42 Squadron had just taken delivery of its first MR.2 aircraft and these were working alongside its earlier MR.1As. Around the same time, interest in the Shackleton was being expressed by the South African Air Force and a visit to South Africa was added to the task. It meant that the six aircraft scheduled to visit Ceylon were split into two flights: 'A' Flight, equipped with three MR.2 aircraft (who would return from Ceylon via South Africa), and 'B' Flight, with three MR.1 aircraft (who would return to St Eval using the reverse of their outbound route).

Following the successful conclusion of Fleetex, the three MR.1 aircraft returned to the UK. Meanwhile, the three MR.2 aircraft routed via Mauritius to South Africa for their goodwill visit to the SAAF at Langebaanweg (Cape Town), before returning via the Belgian Congo, Nigeria and Libya. In just 24 days away from home, the three MR.2 aircraft flew 20,000 miles. Enormous credit, however, is due to the groundcrew whose sterling work enabled each MR.2 to fly an average of 130 hours in a little over three weeks, without once deviating from schedule.

On 12 August 1954, four MR.1 and MR.1A aircraft from 206 Squadron did a seven-week Commonwealth tour, with exercises in Ceylon and goodwill visits to New Zealand and Fiji. No 204 Squadron sent four of their MR.2 aircraft to Durban, South Africa, in June 1955 for Exercise Durbex II with the SAAF and South African Army, flying via the Gold Coast for a goodwill visit on the way home.

In July 1960, four Shackleton MR.2 aircraft of 204 Squadron flew to Singapore via Kindley Field (Bermuda), Palisadoes (Jamaica), Trinidad and Stanley Field (British Honduras). While at Trinidad the squadron demonstrated the versatility of the type by taking troops from the local regiments to Stanley Field for troop rotation utilising a total of 29 trips.

Royal escorts

When HRH Princess Elizabeth and the Duke of Edinburgh flew to Canada in 1951 aboard a BOAC Stratocruiser, G-AKGK *Canopus*, it was the first time royalty had flown

ABOVE While 42 Squadron were deployed to Ceylon in 1953, they sent a small detachment on to South Africa. Shackleton MR.2s of 42 Squadron, including WG554/A-A and WG556/A-J, were photographed while flying over Durban on 22 April 1953. The tour to the Far East and South Africa covered 17,740 miles in April and May that year. *(Crown Copyright/Air Historical Branch image CFP-744)*

BELOW Five Shackleton MR.1As and a single MR.1 from 220 Squadron were photographed at RAF St Eval in June 1952 while practising for the forthcoming Queen's birthday flypast over Buckingham Palace scheduled for later in the month. The aircraft are MR.1As WB831/T-Q, WB837/T-B, WB824/T-O, WB823/T-N and WB828/T-K, while MR.1 VP257/T-P completes the formation. *(Crown Copyright/Air Historical Branch image PRB-1-4663a)*

across the Atlantic. It also became the first of many royal SAR standbys by Shackleton crews. Their Royal Highnesses returned on board the Canadian Pacific liner *Empress of Scotland*, which was intercepted by a 120 Squadron aircraft on 14 November and escorted in mid-Atlantic for 45 minutes while a message of welcome was sent by Aldis lamp – there was no ship–aircraft R/T facility in those days!

On 15 October 1956, 204 Squadron had to fly a little further afield on royal escort duties when WL738 and WL740 were despatched to cover the flight of HRH the Duke of Edinburgh from Gibraltar to Kano, Nigeria.

MOTU

The Maritime Operational Training Unit (MOTU) was formed on 1 October 1956 when 236 OCU was combined with 1 Maritime Reconnaissance Squadron (1 MRS). At the time, it had 22 different MR.1 and MR.1As on strength. Around this time a decision had already been taken regarding the standardisation by Coastal Command on the Shackleton and, as a consequence, they withdrew the remaining Neptune aircraft from service. At the same time the training system was overhauled and the need for a dedicated trainer version of the Shackleton was identified. Special versions of the MR.1A modified to T.4 configuration were ordered, the prototype being VP258.

The first 'new' training aircraft issued to MOTU was VP259, delivered in March 1960. Eventually, all 17 airframes converted from MR.1A to T.4 configuration were operated by MOTU at RAF Kinloss.

In October 1965, MOTU moved to RAF St Mawgan where the remaining T.4 aircraft were replaced with T.2 conversions from MR.2 airframes, the last one being converted by Hawker Siddeley at Langar in 1967. Later, these were replaced with MR.3 aircraft.

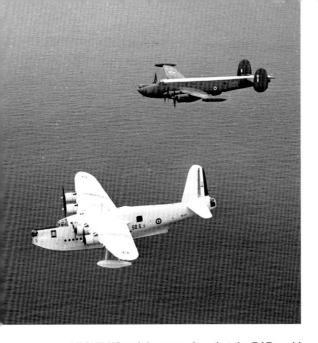

ABOVE When it became clear that the RAF could not preserve a Sunderland – given the lack of staging facilities combined with the significant costs of flying one back to the UK – it emerged that the Aeronavale (French Navy) operated three Sunderland aircraft from Toulon. They were approached and agreed to donate one of their aircraft. Sunderland V, ML824, made its final flight from Lanveoc-Pouloc, near Brest, to Pembroke Dock on 24 March 1961. En route, she rendezvoused with a pair of 201 Squadron Shackletons (from her old squadron) over the Bishop Rock and flew in formation via St David's and Pembroke. The two St Mawgan-based Shackleton MR.3 aircraft were WR975/O and WR980/P. ML824 is now preserved in the RAF Museum at Hendon. *(Crown Copyright/Air Historical Branch image PRB-1-20681)*

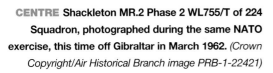

TOP RIGHT An image taken during the NATO Exercise Dawn Breeze held in March 1962 off the Iberian peninsula. A Shackleton MR.3, code 'F', of 206 Squadron, was photographed during a simulated attack on a Royal Navy submarine. *(Crown Copyright/ Air Historical Branch image PRB-1-22420)*

CENTRE Shackleton MR.2 Phase 2 WL755/T of 224 Squadron, photographed during the same NATO exercise, this time off Gibraltar in March 1962. *(Crown Copyright/Air Historical Branch image PRB-1-22421)*

RIGHT Four St Mawgan-based Shackleton MR.3s took part in a formation fly-by over the 50th anniversary celebrations of the Royal Air Force at RAF Upavon in June 1962. The aircraft are WR986/M (leading), XF709/N and XF730/N of 210 Squadron, along with WR985/A of 206 Squadron. *(Crown Copyright/Air Historical Branch image PRB-1-23085)*

ABOVE A 206 Squadron Shackleton MR.3, coded 'B' (unidentified although thought to be WR979), was photographed in 1963, before the Phase 3 modifications were carried out in 1968. At the time, 206 Squadron were based at RAF St Mawgan but moved to RAF Kinloss in July 1965. *(Crown Copyright/ Air Historical Branch image PRBX-1-10133)*

The final Shackleton training course was completed at MOTU on 28 July 1970 and on 1 August it was renamed 230 OCU to undertake crew conversion training for the Nimrod MR.1.

JASS Flight

On the formation of NATO, the United Kingdom assumed a major anti-submarine role across the eastern Atlantic and North Sea areas. During the latter stages of the war an anti-submarine tactics school had been established at HMS *Sea Eagle*, just outside Londonderry, and afterwards this idea was further developed into what became known as the Joint Anti-Submarine School (JASS). Commanded jointly by RN and RAF personnel, JASS was officially opened on 30 January 1947. The unit had its own air elements, Royal Navy Barracudas of 744 Squadron based at Eglinton, with RAF's JASS Flight, based at the then reopened RAF Ballykelly. JASS was initially

equipped with two Lancasters, one Warwick and one Anson. The task at JASS was to run courses to teach tactical doctrine and conduct anti-submarine warfare (ASW) with emphasis on the development and application of combined tactics.

JASS took delivery of its first three Shackleton aircraft to replace the ageing Lancasters in June 1952. JASS Flight provided the 'opposition's' air component, carrying out shadowing of the surface forces on behalf of the submarines during exercises mounted jointly by the Royal Navy and RAF for personnel on the courses.

Each operational Shackleton squadron would spend three weeks every year working with JASS. This involved ground instruction in tactics and techniques, followed by theoretical exercises at HMS *Sea Eagle*, the naval shore establishment in Londonderry. The practical side would then follow involving ships, submarines and aircraft from NATO countries operating in the Atlantic North-West Approaches. At the end of each phase all personnel would return to HMS *Sea Eagle*, find out how well or badly they had done and argue about the outcomes! The object of the exercise was constantly to develop and improve the techniques involved in the combined air–sea approach to anti-submarine warfare, which was vital as the Soviet Union was continually improving and enlarging its submarine force.

At the beginning of 1955, JASS Flight

RIGHT RAF Changi-based 205 Squadron started to take delivery of Shackleton MR.1As in May 1958 and they remained in continuous service until they were replaced by MR.2 examples in September 1962. No 205 Squadron MR.1A, WG525/E, was photographed while visiting Khormaksar in early May 1962. *(Ray Deacon)*

ABOVE Another well-travelled 120 Squadron Shackleton MR.3 was WR989/B, photographed while visiting Eastleigh, Nairobi, in July 1963. *(Ray Deacon)*

RIGHT Avro Shackleton MR.3, XF707/C, of 206 Squadron, in flight over the British coast during a flight from RAF St Mawgan in 1964. This aircraft had undergone its Phase 2 modifications at the time of the photograph and later became the very last Shackleton MR.3 to receive the Viper auxiliary engines when they were installed during 1967–68. *(Crown Copyright/Air Historical Branch image T-4753)*

replaced their Shackleton MR.1s with three new MR.2s – WR969/A, WR967/B and WR966/C. Unfortunately, the unit's life was to be short; at the beginning of March 1957 it was disbanded with WR966 and WR969 being delivered to 220 Squadron and WR967 going to 42 Squadron.

Development work and service trials

Significant amounts of trials work was undertaken with the Shackleton. Some of the highlights include the MAD (Magnetic Anomaly Detector) boom fitted to MR.2, WL789, in an extended tail. The aircraft was converted by a work party from Avro at 38 MU RAF Llandow in June 1953, before the aircraft was moved to St Mawgan where it was placed on the strength of the ASWDU. Despite more than four years of extensive work, the Shackleton proved unsuitable for this equipment due to the high resonance of the engines and contra-rotating propellers. WL789 was later converted back to standard configuration and eventually joined 224 Squadron at RAF Gibraltar.

BELOW A mixed formation of the Examining Wing of the Central Flying School at Little Rissington on 24 July 1953. The Shackleton MR.1A leading the formation is WB861 and is carrying the Ballykelly code 'L-D' of 240 Squadron, so was probably on loan to the CFS at the time of this photograph. Other aircraft in the formation are Vampire T.11, WZ576/35; Meteor T.7, WA630; Varsity T.1, WJ892/Q; Sabre F.4, XB547; and Balliol T.2, WG185/Q-X of 7 FTS. *(Crown Copyright/Air Historical Branch image PRB-1-6724)*

LEFT Shackleton MR.1, VP293, made its first flight on 18 July 1951 and was issued to 236 OCU. In August 1956, it was returned to Avro at Langar for conversion to T.4 configuration. In January 1964 VP293 was purchased by the MoA and transferred to the RAE at Farnborough for trials work, where it remained until withdrawn from use in May 1975. VP293 was photographed at Farnborough in 1968 before receiving the DayGlo paintwork to its nose, fins and rudders, as well as the propeller spinners. (See also page 24). *(Ray Deacon)*

LEFT Shackleton MR.3, XF705/C, of 203 Squadron, photographed with an unidentified pair of Canberra PR.9 aircraft from 13 Squadron while approaching the coastline of Malta in July 1969. Both units were resident at RAF Luqa at the time of the photograph. *(Crown Copyright/Air Historical Branch image TN-1-2510)*

A Lockheed Neptune was used in trials for rocket-fired 'Glow Worm' illumination flares intended to be used on the Shackleton. The trials proved the system to be somewhat hazardous and although the MR.3 aircraft were constructed with the wiring in their wings for operating the rockets, the system was never adopted by Coastal Command.

The Sara airborne lifeboat was fitted to prototype Shackletons and tested. The arrangement was discarded at an early stage in favour of the more efficient Lindholme Gear. Despite these early findings, WL796 was used for trials with a Mk 3 airborne lifeboat and exhibited at the SBAC Show at Farnborough later in the year with the airborne lifeboat under the front fuselage, though the aircraft never actually dropped it. WL796 was later delivered to 38 Squadron in January 1954.

LEFT An unidentified Shackleton MR.3 aircraft coded 'C' (although thought to be XF705) of 203 Squadron, from RAF Luqa, Malta, photographed over the MV *Mary Lou* and its tug in the Mediterranean shortly after the squadron took up residence on the island in February 1969. The Shackletons remained on the island until the type was replaced with the new Nimrod MR.1 in October 1971. *(Crown Copyright/Air Historical Branch image TN-1-2618)*

RIGHT No 205 Squadron had originally moved to RAF Changi – located on the eastern end of Singapore – in March 1958. At the time, the squadron was operating Sunderland GR.5 aircraft. These were replaced by Shackleton MR.1A aircraft beginning in May 1958 and then subsequently replaced by the MR.2 variant in February 1962. No 205 Squadron remained at RAF Changi until disbanded in October 1971. WR959/A of 205 Squadron was photographed at Changi in August 1968. *(Peter R. March)*

RIGHT WG532/E of 205 Squadron photographed during a sortie from RAF Changi in August 1968. WG532 was ferried to 27 MU at Shawbury and declared a non-effective aircraft (NEA) shortly after this image was taken. *(Peter R. March)*

Institute of Aviation Medicine

Although the Shackleton MR.1, MR.1A and MR.2 had proved themselves to be excellent maritime reconnaissance aircraft, the crew working conditions left something to be desired. As a result of a number of complaints regarding crew fatigue, trials were undertaken in 1953 and 1954 by the Institute of Aviation Medicine (IAM) using crews from 240 Squadron at RAF Ballykelly.

This trial required the crews to undertake 60 hours of night flying over a seven-day period. Each flight took off at between 6.00pm and 6.30pm each day with the crews flying right through the night and landing around 7.00am

RIGHT Avro Shackleton MR.3, XF700/F, of the resident 203 Squadron, taking off from RAF Luqa, Malta, during Exercise Lime Jug in October 1970. The aim of the exercise was to practise procedures for co-operation between Royal Navy forces and RAF shore-based aircraft in the Mediterranean against the perceived Soviet threat. The exercise ran from 2 to 15 November and, in addition to the resident Shackleton MR.3s, also involved Lightning F.6, Buccaneer, Phantom FG.1, Nimrod MR.1, Canberra T.17 and Victor K.1 aircraft of the RAF, as well as Royal Navy Sea King and Buccaneer aircraft. *(Crown Copyright/Air Historical Branch image TN-1-6293-143)*

the next morning. Each crew member was monitored throughout the flights to determine the impact of the aircraft's characteristics on their own abilities.

The trials indicated that each and every one of them was affected in one way or another, the most noticeable impact being a continual buzzing in their ears from the propellers. Others suffered a physical weight loss of several pounds after each sortie. All of the factors identified highlighted the need for change. While more Shackletons were required to replace the remaining Sunderland aircraft, production of the MR.2 was stopped in September 1954.

A complete redesign was required, with a list of improvements in many areas, which effectively resulted in the design and delivery of the Shackleton MR.3.

Christmas drops to weather ships and fisheries protection vessels

The first mail drop to an ocean weather ship (OWS) by a Shackleton was conducted in December 1951. The vessel – the *Ocean Explorer* – was operating on station 'JIG' located at 5230N 2000W. It received no fewer than three visits during the month – including the Shackleton OCU dropping a Christmas tree. A BBC television camera was on board the Shackleton but, sadly, no one at RAF Kinloss saw the images as the station was well out of range of the nearest television transmitter.

For the Ballykelly-based Shackleton squadrons, the use of the ocean weather ships for homing practice or as waypoints in navigational exercises was commonplace, so they were delighted to drop the Christmas goodies to the vessel located at 'Station Juliet' just a few days before festivities, a commitment which endured until the ship was withdrawn from service in the 1960s.

Squadron Leader (Retired) Mike Rankin recalls with affection a Christmas drop for a fishery protection vessel in November 1958:

The task was to drop Christmas mail to the fisheries protection frigate HMS Duncan off the north coast of Norway. We were given the ship's PIM (position and intended movement). She was between North Cape and Bear Island and we would have to intercept her at the middle of the day to maximise the light available so far north. Checking up on the daylight times, the navigators found that official dusk occurred just 15 minutes after official dawn on 11 November! They calculated back from the centre of the 15 minutes of daylight, that we would have to leave RAF Ballykelly at 01:30 GMT. We would have only short range VHF communication with her.

As usual, we were flying under radio silence on short wave, so it was with some misgivings that we viewed the prospect of making contact with Duncan early enough to get us into position to drop her the goodies in daylight. It not only required us to be very accurate, but it also required Duncan to make good the PIM on which our navigation had been based.

When the time came to begin our search for her, the radar was switched on and – miracle of miracles – there was a big fat radar contact exactly where we wanted her and we arrived with spot-on timing. The light was good as we began the run-in and dropped the packages, so I flew round again to take her picture. But by the time we were in a place to do it, there was too little light available to make it possible. However, it did explain very plainly why we were dropping Christmas goodies at least six weeks before Christmas.

We stayed for twenty minutes or so while they told us of some of the hardships of operating in such high latitudes. Worse was to come, it seemed, since they were to leave this area and perform their duties in the South Iceland Sea area over Christmas. Late in 1959 they were going into the heart of the then Cod War, which had always had the potential for violent confrontation. A few minutes after we had left them for our destination at Bodø, there came a loud squealing noise, suggesting a short circuit in the intercom system, but it was quickly found that the noise was on the VHF radio, to which we had all been listening. Very soon someone recognised the sound of singing.

R 121110Z

FM HQ 18 GP

TO RPDLKB/BALLYKELLY

BT

R E S T D AT 241 NOV 12

FOR ATTENTION OC 210 SQN PD FOLLOWING RECEIVED FROM DUNCAN

QUOTE AIR DROP MOST SUCCESSFULLY CARRIED OUT IN POSITION

7305N 1810E IN PERFECT WEATHER CONDITIONS PD WE ARE MOST

GRATEFUL PD HYMNS ANCIENT AND MODERN 224 LAST VERSE IS

APPOSITE UNQUOTE

R

ROUTINE

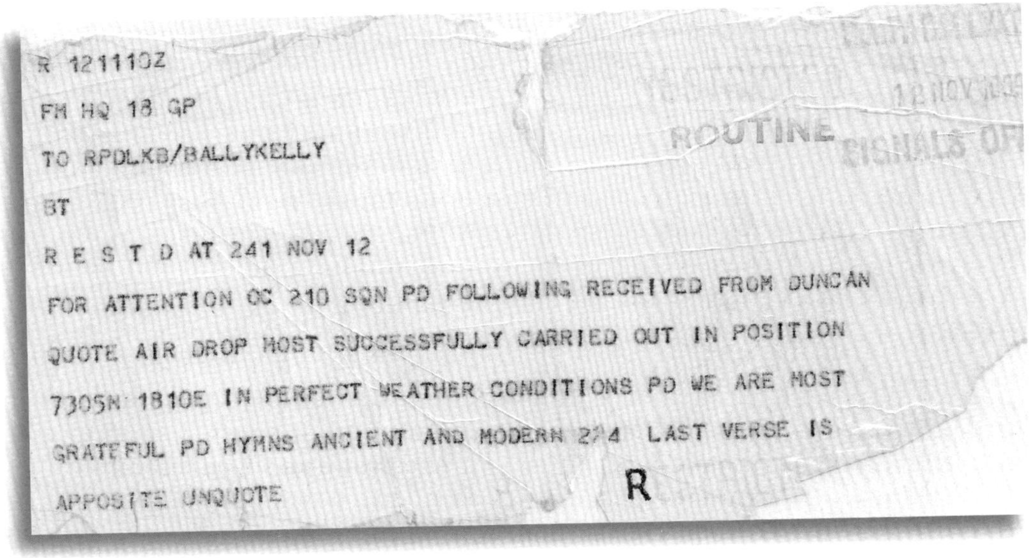

LEFT A copy of the signal from HMS *Duncan* via the MHQ in Scotland. *(via Squadron Leader Mike Rankin)*

Try as we might we could not work out either the tune, or the words. Since we had been commiserating with them on their prospects for a fun Christmas, we assumed that it must be a Christmas carol. So we gave them an expert rendering of the first verse of 'Good King Wenceslas', which everyone thought he could remember.

Back at Ballykelly two days later I was called to the Squadron Commander's office to be told that he had received a copy of a signal from HMS Duncan *via our MHQ (Military HQ) in Scotland. A scanned copy of what is left of the original signal is shown above.*

It refers to the last verse of Hymn 224 in Hymns Ancient and Modern*, which reads:*

O happy band of pilgrims
Look upwards to the skies
Where such a light affliction
Brings forth so great a prize.

It was a nice way of expressing their gratitude for what was obviously an important boost to their morale, so far away from home.

Search and rescue

Twenty-four-hour standbys commenced in 1952, each squadron taking a week in turn. 'Scrambles' soon followed and on 22 February, a 120 Squadron Shackleton escorted a USAF C-54 that had suffered an engine failure over the Atlantic. This was the first of many such incidents, some of which proved rather embarrassing to the Shackleton because some airliners could fly with one engine shut down and still outpace its rescuer! The most frequent customer of the time was the Lockheed Constellation, which had notoriously unreliable Wright Duplex Cyclone engines, but had no problem in outpacing its rescuer, so a successful interception required careful planning.

In their heyday Shackletons were frequently able to assist stranded mariners by dropping equipment, particularly Lindholme Gear, which soon after the Second World War was found to be superior to the airborne lifeboat. Lindholme Gear, named after the South Yorkshire RAF station where it was invented, consists of three units linked by floating rope giving a spacing of around 200 yards between the first and second unit, and 400 yards between the centre and third. The large centre unit consisted of an MS9 life raft, capable of accommodating up to nine people with numerous survival aids aboard, including fresh water, food, first-aid equipment, a homing beacon, tropical or Arctic fishing equipment, a de-salter and such morale-boosters as cigarettes, seasickness pills and playing cards. The two smaller containers were packed with other stores, some of them dependent on climate.

The object was to drop the gear in such a way that the floating ropes wrapped themselves around the survivors so that little or no effort was required on their part. This called for precise flying at low level, careful planning and frequent practice. Final drops were made at speeds between 140 and 160 knots.

Oceanic survey work

As part of the International Geophysical Year of 1956, 204 Squadron took part in ocean survey work throughout the summer.

ASWDU

The Air/Sea Warfare Development Unit (ASWDU) had moved up from RAF Thorney Island to Ballykelly in March 1947. It received its first Shackleton MR.1 aircraft (VP261) on 27 April 1951. The ASWDU had moved to Ballykelly when the airfield was reopened as a Coastal Command base for ASWDU to carry out trials of new electronic equipment, in order to ascertain the suitability of the equipment for general use in Coastal Command and, in particular, for maritime reconnaissance duties.

However, its stay at Ballykelly was relatively short-lived and it moved to RAF St Mawgan on 10 May 1951. Soon afterwards, it received MR.1, VP282.

With the much-improved MR.2 becoming available, ASWDU was one of the first units to receive that version, accepting WG532 for use in under-wing rocket trials in January 1953. These proved unsuccessful and rockets were never adopted for use with the Shackleton.

As mentioned earlier, MR.2 WL789 was modified with MAD equipment and joined ASWDU on 10 September 1953. However, despite 4½ years of extensive trials, the system did not perform on board the Shackleton.

On 1 September 1958, ASWDU moved back to Ballykelly from St Mawgan after an absence of more than 7 years, bringing three aircraft with it. The unit tended to rotate its aircraft, dependent upon what trials were being conducted. Over the next couple of years it used examples of both early marks of Shackleton. One of the most important tasks being undertaken around the time of the move was the operational evaluation of ECM equipment, known as Orange Harvest, which would become part of the Phase 2 package of improvements.

The unit was different from the others at Ballykelly and worked very hard at trying to maintain those differences. First of all, it reported directly to Coastal Command Headquarters

BELOW The scene at NAS Key West, Florida, during the first annual conference on anti-submarine warfare development, 1966. In addition to a Shackleton from the ASWDU, the event was also attended by a team from the RCAF with an Argus (left), along with a US Navy P-3 Orion from VX-1 Squadron. *(via Squadron Leader Mike Rankin)*

at Northwood and not to 18 Group as the operational squadrons did. This meant that it was a 'lodger' unit and could operate in relative isolation from the rest of the station – thereby not participating in orderly or SAR duties.

ASWDU continued its development work on the Shackleton for many years and also co-operated with comparable units operated by the RCAF (the Maritime Proving and Evaluation Unit at Summerside) and the US Navy (VX-1 Squadron). Indeed, more often than not, aircrew from these units were on exchange postings with ASWDU, and vice versa.

In 1966, the first annual conference on anti-submarine warfare development was held at NAS Key West in Florida and was attended by the UK, the USA and Canada. ASWDU took a single MR.2 while the RCAF had a Canadair Argus present, alongside a US Navy P-3 Orion.

The ASWDU disbanded on 1 April 1970, when control of all trials work for the Nimrod fleet was assumed by the Central Trials and Tactical Organisation.

Buckets of sunshine

The first use of Shackletons in support of nuclear testing took place when four modified MR.1s of 269 Squadron were detached from their Ballykelly base to Darwin, Australia. While there, the aircraft undertook various weather data-gathering operations over the Timor Sea and Indian Ocean. Detachments to Alice Springs saw the Shackletons performing meteorological reconnaissance, in addition to patrolling those areas designated as

RIGHT A view of the base at Christmas Island, taken just a few days before the final thermonuclear test. Visible here are 9 Shackletons – the other may well have been taking this photograph – and 11 Canberra aircraft, some of which would have been from the RAAF's 76 Squadron in South Australia. All the Canberra aircraft would have been involved in sniffing at the contents of the cloud immediately after each explosion, or flying the filter papers on which the samples had been trapped, back to the Atomic Weapons Research Establishment in the UK. *(via Squadron Leader Mike Rankin)*

being 'dangerous' during the tests, to ensure they were kept clear of all shipping. All of these sorties were completed under the code name Operation Mosaic and, once completed, the aircraft returned to the UK.

Following agreement with the Australian government, it was planned to detonate a limited-yield thermonuclear device at the Maralinga Range, under the code name Operation Buffalo. Four Shackleton MR.2 aircraft from 204 Squadron were detached to Pearce AFB in Western Australia. The first British thermonuclear device (Blue Danube) was dropped over the range by Valiant WZ366 on 11 October 1956. Having completed the testing, the Shackletons returned to the UK in November.

Permission for further testing was declined by the Australian government and as a consequence two Shackletons of 206 Squadron were tasked with circling the world westwards as a preliminary to establishing Christmas Island as a base for further proposed thermonuclear tests in the area. After an initial survey, Christmas Island was indeed deemed suitable for the tests, although the small matter of constructing two runways, along with other infrastructure still had to be overcome.

Aircraft from 206 and 240 Squadrons at Ballykelly were selected to support the trials. Firstly, the aircraft had to be specially modified to partake in the testing and were despatched to 49 MU for the work to be completed. The first detachment of 240 Squadron took place at the end of February 1957. In April 1957, the aircraft returned to Ballykelly after the successful drop of the first weapon in Operation Grapple 1.

In January 1958, 240 Squadron was to return to Christmas Island in support of the fifth H-bomb test, under the code name Operation Grapple Y, with a planned return of 3 June. Meanwhile, their place on Grapple Y had been taken by 204 Squadron, who flew out to the island in May, where they also stayed for the sixth H-bomb tests, Grapple Z. In June 1958, they were joined by a detachment from 269 Squadron and once the final bomb had been detonated – on 11 September 1958 – all aircraft returned home, arriving back at Ballykelly in October.

OPERATION GRAPPLE REMINISCENCES

Squadron Leader Mike Rankin was a young co-pilot with 269 Squadron when he and his crew were instructed to join the Christmas Island detachment. He recalls:

The Shackletons' long range and endurance made possible meteorological reconnaissance from as much as a 1,000 nautical miles' radius from the island. The ten MR.1 aircraft and their crews provided by 204 and 269 Squadrons from Ballykelly also permitted a 24-hour Search and Rescue facility to be set up for the duration of the detachment.

Prior to each test, a full rehearsal of the entire operation was carried out. Weather investigations were carried out for days before the test and during the whole of the night before a test; all ten aircraft were engaged in coordinated searches to ensure that the entire published danger area was free of intruders or potential accident victims.

My impressions of the flying at Christmas Island were heavily coloured by the youthful enjoyment I still found in being the co-pilot of a Shackleton crew during my first couple of months and with so much to learn. Everything was new to me, including the beautiful location so close to the Equator in the staggeringly large Pacific Ocean and the crew I had joined so recently. The crew quickly accepted that I did at least know how to fly the aircraft and, after the flight engineer had casually treated me to a long discussion of the technical details of the Shackleton, that I probably knew enough about the aircraft for them to relax.

There were two memorable flights in connection with the nuclear tests. The rehearsal of a search immediately before the first test was the first time I had ever been required to stay awake all night and I did not manage to do it. Daytime rest at Christmas Island was not easy to achieve, especially if you were sharing a tent with two guys not on your own crew. Soon after dark it was my first stint in the left seat and I quickly fell asleep, unaware that it was creeping up on me. My captain, in the right seat, which in the Shackleton is separated from the left by a passageway to the nose compartment, kicked me hard in the right thigh across that passageway. It was a shock. By now I had sufficient incentive and managed to remain awake for the rest of that long night rehearsal.

A few nights later, the first test was made with one of the triggers. The search we had been allocated was one of the most distant, which meant that when the trigger was detonated around 9.00am, we saw nothing of it.

For the first test of an H-bomb our crew was stood down, and we went through a very different procedure designed to ensure that everyone on the island was known to be in a safe area. We rose very early, breakfasted and assembled in a large clear space on concrete where we were counted and kept in locations marked out for our squadron. Altogether there were approximately 7,500 people on the island and those who were not flying had to be accounted for. It was a long, tedious process for our own safety, leaving all domestic, engineering and office spaces completely empty and quiet everywhere. We were briefed on what was to happen and finally we all sat facing away from 'Ground Zero', 22 nautical miles behind us, just off the southern tip of this giant atoll.

We were made aware of the take-off of the Valiant that would launch the weapon, and once it was in position for the drop, the public address system was connected, not only to the communications channel with the aircraft but also to the crew's intercom. To cap this extraordinarily intimate connection, the Valiant was dragging a persistent condensation trail behind it and was clearly visible. The commentator had an exceptionally good speaking voice and manner, resulting in a 3D audio-visual display never equalled by any television show that I have ever seen. The real drama was when the Valiant turned on to its attack run from the north-east, its track pointing directly at us. We were directly between the aircraft and its target!

We listened to the crew's conversation during the run-in as appropriate switching was selected and the launching system was proved. There was a final 10 seconds or so of complete silence from both the public address system and the many hundreds of spectators until the bomb was released. I think most of us had been concentrating very hard on the movements of the aircraft and it was only now that we realised that the bomb had actually been released while it was a considerable distance in front of us. We were obliged to clasp our hands together over our eyes to protect them from the flash of the explosion at this point. It finally struck home that the bomb was flying over our heads and had a considerable distance to fly before it met its end at 7,000ft above the sea. It seemed a very long fall, which was accompanied by a long, continuous tone from the public address system that would be interrupted immediately before the explosion. It would remain silent for 30 seconds, after which time we could uncover our eyes and turn round to see the result that would be presented in the sky above the coconut forest surrounding us.

We were all conscious of the flash, firstly from seeing the glare through our hands, then quickly from the heat on our backs. I felt it through my flying suit. It grew hot and hotter; and kept increasing until I began to wonder whether someone had miscalculated and we were about to be fried. Thankfully, nobody had miscalculated, but for a couple of days afterwards my back itched, no doubt for psychosomatic reasons.

When we were cleared to turn around, the part of the cloud visible to us had already cooled below red heat and was now a scarcely believable mass of dark colour, with areas of charcoal, purple and brown predominating, the whole growing rapidly and rolling in violent internal turbulence. The double bang, when it came, was not very loud but we could see it was coming as it travelled past and alarmed the many frigate birds, terns, common gulls, gannets and other unidentified species that recoiled visibly and dived seawards for safety. We would later see seriously injured birds, mostly blinded and to a lesser extent burned. I know that our groundcrew, not normally known for being soft-hearted, collected some of those unable to fly and tried to help them. They had a great disappointment waiting for them when their efforts to revive the birds failed, without exception.

For some minutes after the double bang, the cloud's extreme turbulence could be heard as a continuous roaring from the sky. The extent of the cloud was immense – perhaps as much as 90° from left to right. The combination of the continuous noise and ugly colours, allied with the violent turbulence never seen even in the worst cumulonimbus storm, was unforgettably menacing. There was very little talk about it in the mess, ever. It had had a very sobering effect on us all. We chose to keep our thoughts and memories of that morning entirely to ourselves.

Then there was one last weapon to be tested. Of the ten-aircraft search plans, ours was to be placed in the centre, clearing the waters immediately round the atoll. We took off last, in recognition of the shortest transit distance and to avoid mixing it with the other nine as they left. When our search had been completed, we had another job as we were carrying a photographer employed by the Atomic Weapons Research Establishment (AWRE) to film the explosion and development of the subsequent cloud. We were to film it from a distance of 50 miles and flying at a height of 7,000ft, which was the height at which the weapon would explode. That height, we had been told, was just sufficient to keep the fireball clear of the surface of the sea below it. Unusually, our pre-detonation search height was also 7,000ft, which gave our ASV-13 radar a better range, simply by extending the radar horizon.

In the early hours just before dawn the radar picked up a target, very close to ground zero. I was again at the controls and was surprised to be told not to descend to identify the target. No reason was given, which was also unusual, but I did not ask the obvious question and we flew towards the contact,

which proved to be a naval vessel making itself easily visible by the high-speed wake, which could be seen for many miles thanks to the phosphorescence often seen in tropical oceans.

The professional attitude required of everyone participating in nuclear testing allowed for absolutely no variation from the exceptionally complex and intertwined operation orders, under which all forces operated. Accordingly, we had reported the radar contact immediately. I have often wondered since what was done about this unusual event. The ship was the New Zealand Navy frigate *Otago*. We established contact and were told that she also had a small radar contact, even closer to ground zero. The thought was that it could have been a submarine. They must have had very good radar capabilities because they found the contact, which was a floating barrel and were now heading out of the defined danger area at maximum speed. Our navigator had already plotted the ship's speed and the distance to the edge of the danger area and found that they would just about make it. It was clear that the Operations Centre had not been told what *Otago* had been doing. I suspect that the *Otago*'s captain may have made his last voyage, but of course we never heard any more about it.

We maintained our surveillance of the central search area until it became time to set up our photography task. The sun rose on time. The weather remained as good as it had ever since we had arrived and eventually we found ourselves moving more-or-less eastwards around a large circle of 50 nautical miles' diameter, and with the auto pilot engaged. The weapon detonated exactly where it was intended, with the camera rolling. The crew were now wearing black masks and the pilots also had their hands folded together over their eyes. As with the previous H-bomb explosion, the glare of the flash made itself visible through both black mask and folded hands. When the warning tone sounded again allowing us to look – 30 seconds after the detonation – the flash was still so strong that I waited for a further 3 minutes before my vision was fully restored.

I also remember seeing a volume near the base of the now-rising cloud that was still glowing red hot.

At a 50-mile radius, the cloud did not take up as big a fraction of our view as had the previous H-bomb explosion seen from the ground. We could neither see the turbulence, nor of course hear the roaring of the cloud. But what we did see was one of the most beautiful sights I ever saw in the sky. The visibility was virtually unlimited. By now the mushroom and its stalk were pristine white, the mushroom head visibly climbing. Above the curved surface of the head a very fine line of featureless cirrus cloud was forming as the air above it was forced up. Featureless as it was, at the edges, where it began to fade, it was easily possible to see that this cloud was streaming outwards and cascading down beside the rising mushroom. The mushroom stalk, obeying some unknown law, looked as though unseen hands had grasped the middle of the stalk and rammed the lower part upwards into it, widening the base of the assaulted upper part. Two more such discontinuities existed further up the stalk as the head climbed into the stratosphere. Writing this description is a poor substitute, but my mind still holds that image of real beauty and will do so until I die.

Much as I enjoyed all the flying I did at Christmas Island, it was relatively simple. However, back at Ballykelly, while we were undergoing conversion from the Mk.1 to the Mk.2, my captain suddenly asked, 'Did you ever wonder why I wouldn't let you descend when we first picked up that contact on the frigate *Otago*?' He continued that, at the time, he had not been allowed to tell me but he could see no reason for withholding that information now, especially since there was no further prospect of nuclear tests that needed our support. It had been part of a measure to minimise the chaos that would have occurred had the Valiant bomber crashed on take-off AND if its H-bomb had also exploded. 'How did that affect us?' I asked. I mentioned earlier that we had been the last of our ten aircraft to take off and gave some reasonable justifications for that late departure. But it turned out that we needed to conserve fuel

ABOVE **The result of the final thermonuclear test carried out at Christmas Island on 15 July 1957, as seen by the AWRE photographer on board a Shackleton aircraft.** *(via Squadron Leader Mike Rankin)*

as far as possible because, in the event of that unthinkable (and very unlikely accident), ours would have been the only one of the ten aircraft that had enough fuel to reach another usable airfield!

Descending from 7,000ft to our more usual 1,000ft and then having to climb back up again for the photography would have necessitated using 200 or more gallons of fuel that might have entirely negated the precautions already taken to ensure that at least someone would be able to tell the story of Britain's final atmospheric nuclear weapons tests!

Chapter Three

Shackleton at war

Having originally been designed to fulfil the maritime reconnaissance, anti-submarine and air-sea rescue roles, the versatile Shackleton was called upon to demonstrate its other capabilities in both colonial policing and as a conventional bombing aircraft. During its RAF career, the Shackleton – in its various marks – operated in a variety of roles within a number of theatres.

OPPOSITE An Avro Shackleton MR.2C, WL738/D of 37 Squadron, flying over the Federation of Southern Arabia (now Yemen) during operations in the Radfan region of the country in 1964. The squadron was based at RAF Khormaksar, Aden, at the time. *(Crown Copyright/Air Historical Branch image T-4646)*

The Cyprus emergency

Having been subject to Turkish sovereignty until the First World War, Cyprus was annexed by Britain in 1914 and became a colony in 1925. With 80% of the population being of Greek origin, pressure grew for amalgamation with Greece. Recognising the difficulty of getting the British to concede such an important base, the National Organisation of Cypriot Fighters (EOKA), under the leadership of the ex-Greek Colonel George Grivas, resorted to arms.

This worsening political situation in Cyprus created a new role for the Shackleton – anti-smuggling patrols to prevent arms reaching the island. No 38 Squadron, based at RAF Luqa, were giving the task of patrolling the coast of Cyprus and flew the first operation on 21 July 1955. It was to become a major undertaking for the squadron, involving 250 flying hours per month and eventually going on to last 4 years. No 42 Squadron assisted in 1957 – during the disturbances – and the last of the 884 sorties was flown by 38 Squadron on 14 December 1959.

On 1 April 1955 a bombing campaign started with attacks on government buildings at Larnaca, Limassol and Nicosia. There was already a large British Army presence following the departure from Egypt in 1954 but, after further attacks in the autumn in which policemen and servicemen were killed, the Governor of Cyprus declared a state of emergency on 27 November 1955. The number of troops on the island was immediately increased under Exercise Encompass. From

December 1955 through to 24 January 1956, this exercise saw numerous Shackletons, from just about every unit that operated the type, being used as troop carriers. The Shackletons continued to serve in this troop-carrying role until the emergency ended in December 1959.

Squadron Leader Mike Rankin was involved in some of the trooping flights towards the end of the campaign. He recalls:

After completing my Shackleton training I was posted to my first 'tour' – with 269 Squadron at Ballykelly in June 1958. Despite achieving over 1,000 hours to date, I had little more than 125 hours on the Shackleton. Thankfully, 269 Squadron was equipped with the MR.1 on which I had been trained and my introduction to the squadron was, therefore, relatively straightforward. I became co-pilot to the 'A' Flight Commander and I began picking up new experiences at an astonishing rate.

Of Squadron Leader Brian Robinson's ten-man crew, only he and I were commissioned. As our captain also held a busy appointment, I was left to look after the crew in his absences, which allowed me to get to know the crew quickly. They were all very experienced in maritime operations so I trod very carefully while we gathered information about one another.

My third flight was from Ballykelly to Abingdon, to take part in an operation to reinforce the Army in Cyprus. We had left three signallers and one navigator behind. At Abingdon we attached a number of slings to

fittings in the floor and, when it was our turn, 31 soldiers joined the now six-man crew for the flight, initially to RAF Luqa, Malta. Their equipment was carried in panniers loaded into the bomb bay. The aircraft had 11 seats in total, so 5 of the most senior men found very comfortable seats available to them and the remaining 20 sat on the bare aluminium floor in those slings, which could prevent them from rolling forward in the event of strong deceleration, but had no way of holding them down in severe turbulence. They were amazingly patient for the 6-hour 45-minute flight to Malta, where we left them to be taken on to Nicosia. After a 12-hour break at Malta we picked up another 31 and took them onwards to Nicosia with a further 5-hour flight.

Our crew then returned to Ballykelly, taking no further part in that large operation. Our squadron's aircraft were needed back at base to prepare them for our impending transit to Christmas Island in support of Operation Grapple Zulu, Britain's final atmospheric test of its newly developed hydrogen bomb.

Suez

The troop-carrying role pioneered by the Shackletons during the Cyprus emergency were to have significant benefit during the Suez crisis when the capability was put to full operational test in Operation Challenger as trouble flared in the region – providing the Shackleton with its first *real* combat

deployment. When President Nasser of Egypt declared that he intended to nationalise the French- and British-controlled Universal Suez Canal Company, the region was thrown into doubt. Fortunately, both the British and French governments has anticipated President Nasser's actions and already had a plan in place. Operation Musketeer was activated while announcements were made from both London and Paris on 30 October 1956.

The Shackleton aircraft in the area were immediately called upon to undertake maritime patrols in support of the operation, but once again they were used in the troop-carrying role. Operation Challenger was put into effect to uplift troops from the UK in support of Operation Musketeer. Five Shackleton MR.1s and MR.1As from 206 Squadron at St Eval uplifted the 16th Parachute Brigade from Blackbushe and delivered them to Cyprus. For the operation, British aircraft had yellow and black stripes painted on to the wings and fuselage, although not all of the Shackleton aircraft had these markings applied.

Nos 37 and 38 Squadrons – from their base at RAF Luqa, Malta – undertook reconnaissance of Egypt from September 1956, as well as anti-submarine protection for the invasion fleet during the landings.

Political pressure for a ceasefire had come from across the globe. A run on sterling resulted in the need for a loan from the International Monetary Fund, in effect dominated by American interests. The Americans agreed to support a loan for $500 million – against a ceasefire. By 6 November the whole crisis was

ABOVE 37 Squadron, based at RAF Luqa, Malta, undertook reconnaissance of Egypt from September 1956, as well as anti-submarine protection for the invasion fleet during the Suez landings. For the operation, British aircraft had yellow and black stripes applied to the wings and fuselage. While not all of the Shackleton aircraft had these markings applied, WL785/E certainly did and is shown with the fuselage markings apparent. (Colin McKeeman)

at an end. The United Nations forces occupied the ground that had been under the control of the British and French troops. The Shackleton was once again called upon to transport the troops out of theatre and back to the UK, with 204 and 228 Squadrons joining 206 in the task.

Malaya

The Malayan emergency was a Malayan guerrilla war fought between Commonwealth forces and the Malayan National Liberation Army (MNLA), the military arm of the Malayan Communist Party (MCP), from 1948 to 1960. The MNLA commonly employed guerrilla tactics: sabotaging installations, attacking rubber plantations and destroying transportation and infrastructure.

The British and Commonwealth operation became known as Operation Firedog. Among a very large number of aircraft deployed into the region, Sunderland GR.5 aircraft from 209, 205 and 88 Squadrons from Changi were detached to Seletar where they were employed in maritime reconnaissance activities, monitoring the seas for vessels involved in arms smuggling. In May 1958, 205 Squadron exchanged their Sunderland aircraft for Shackleton MR.1s and these were soon employed in maritime reconnaissance duties alongside the remaining Sunderland aircraft.

The end of the Malayan emergency was declared from 31 July 1960 and from a British perspective, the campaign was a complete success, though many hard lessons had been learned.

BELOW Avro Shackleton MR.2, WL737/D of 42 Squadron, at Sharjah in July 1957. The Shackletons carried fragmentation bombs and were used during RAF operations against the Omani Liberation Army (OLA) in Oman. The top of the fuselage was painted white in an attempt to reflect the heat of the sun away from the crew members in the fuselage. *(Crown Copyright/Air Historical Branch image CMP-909)*

Oman

Britain has long maintained close links with Oman through a succession of treaties to protect her oil interests and sea routes. For many years Saudi Arabia had contested the border at the crossroads of Buraimi Oasis, but when the area was felt to have significant oil reserves, a Saudi Arabian party of around 80 settled in the village of Hamasa on 31 August 1952. The British government protested, and from 15 September a small force of Trucial Oman Levies (TOL) was despatched to the village. Meanwhile, three Vampire FB.5s of 6 Squadron, supported by a Valetta, were flown to Sharjah from Habbaniyah. After demonstrations and leaflet drops, talks began, and the aircraft returned to Iraq in October.

The talks dragged on and it was decided to mount a blockade of the Saudi enclave. Initially Vampires and Meteors were deployed but these damaged the runway at Habbaniyah so badly that Lancaster GR.3 aircraft from 37 and 38 Squadrons at Malta were deployed, to operate from Sharjah. The aircraft were employed on reconnaissance duties but could barely be spared from their NATO commitments in the Mediterranean and were soon replaced by a pair of Valetta aircraft from the Aden Communications Flight until a pair of Lancaster PR.1s of 683 Squadron arrived. When 683 Squadron was disbanded, these were replaced by Anson aircraft from 1417 Flight until 15 August 1954 when the Saudi Arabian party withdrew.

Throughout the operation the British government were reticent to use offensive airpower as the Saudi party were almost certainly supported by US interests, particularly the Arabian American Oil Company (ARAMCO), which at the time operated C-46 transport aircraft, but also B-26 light bombers.

In January 1955, there were reports of a Saudi Arabian group entering Oman in northern Dhofar, and Valettas, assisted by a pair of Lincoln B.2s of 7 Squadron, mounted reconnaissance sorties. However, these were ended in May that year.

Another incursion by Saudi Arabian 'civilians' at the airstrip at Buraimi in September 1955 was responded to with a force of TOL

supported by Lincolns, Valettas, Ansons, Pembrokes and Venom FB.1s. All of the 'civilians' were removed in Ansons and Pembrokes to Bahrain.

Following further skirmishes in late 1955 and through 1956, a further landing was made near Muscat and within a month the Omani Liberation Army (OLA) had occupied many villages in the area and with the potential loss of central Oman, the sultan sought British help. A pair of 37 Squadron Shackleton MR.2s from Khormaksar was part of the retaliatory force deployed to Sharjah and Bahrain. The Shackletons were also employed on leaflet-dropping duties ahead of major raids. The consequence of all the activity was that the OLA withdrew to the impenetrable Jebel Akhdar, a fertile plateau above 7,000ft with few easily guardable approaches. As a consequence, most British forces withdrew.

Towards the end of 1957, the OLA took the initiative and almost captured Tanuf. The Sultan's Armed Forces (SAF), with RAF help, attempted to oppose them, but this failed. Instead, a blockade was tried to stem the flow of arms. In February 1958, two 17,000yd-range 5.5in Howitzer guns were set up and began shelling the summit villages, while the RAF concentrated on attacking the water supply in the hope of disrupting the pattern of life so that the villagers would deny support to the OLA. The plan appeared to be failing, leading the RAF to step up its bombing campaign.

In one week in September 1958, Shackleton MR.2 aircraft from 37, 42 and 228 Squadrons dropped a total of 148 1,000lb bombs, while

Sea Hawks and Sea Venoms from HMS *Bulwark* joined the action for a few days. Finally, 22 SAS Regiment were brought in from Malaya in preparation for an assault on the plateau in early 1959. Further Shackletons from 224 Squadron arrived in early 1959 – replacing those from 42 and 224 Squadron – and remained until the end of February.

A final assault was made on the night of 26/27 January and the plateau was reached and retaken. Shortly afterwards, most British forces were withdrawn, with only the 152 Squadron Pembroke C.1s and Twin Pioneer CC.1s remaining.

Help for Jordan

From 1955, the Soviet Union provided military aid to both Egypt and Syria, including the supply of aircraft and the building of airfields in Syria. These airfields, located in the west of the country, clearly threatened the integrity of both Lebanon and Jordan. In February 1958, the United Arab Republic (UAR) was formed between Egypt and Syria. Simultaneously, Iraq and Jordan agreed to an anti-communist, anti-Nasser Federation. The situation at the eastern end of the Mediterranean was made even worse with a revolt in Lebanon.

Following the assassination of the Iraqi Prime Minister Nuri al-Said on 15 July, King Hussein of Jordan appealed to Britain the following day for assistance in maintaining stability.

The request was immediately supported and on the next morning 200 troops were moved

BELOW Another 42 Squadron Shackleton MR.2, WR951/E, pictured at Sharjah in the Trucial States during operations against the OLA on the Yemen–Aden border in 1957. *(Crown Copyright/Air Historical Branch image CMP-912)*

to Amman from Cyprus by Hastings aircraft of 70 Squadron. By 18 July, 2,200 troops were in Amman with light artillery support. Reinforcements had been flown into Cyprus by Comet C.2s of 216 Squadron, assisted by Shackleton MR.2 aircraft, once again in trooping configuration (31 soldiers per aircraft) and drawn from 42 and 204 Squadrons. Beverley C.1 aircraft flew in heavy equipment from Cyprus. The troops were followed by a detachment of Hunter F.6 aircraft from 208 Squadron from Akrotiri.

King Hussein established a pledge of loyalty from the powerful Bedouin tribes on 11 August and British troops began withdrawing after a UN resolution called for an end to Western intervention later in the month. The last British troops left on 2 November 1958.

Kuwait emergency

Britain reached an exclusive agreement for protection with the ruler of Kuwait way back in 1899, well before oil was discovered in the 1930s. By 1960 the agreement was somewhat dated and an Exchange of Notes was signed in June 1961 agreeing, among other things, that Her Majesty's government would assist the ruler if requested.

In 1961, around 40% of the United Kingdom's oil came from Kuwait. It was not surprising, therefore, that when on 25 June General Abdul Qarim Kassem declared Kuwait a part of Iraq, the British government reacted sharply.

Intelligence reports confirmed the move of Iraqi troops along the border. The following day all UK forces in the Middle East were placed on standby. At the time, there were three frigates in theatre along with the commando carrier HMS *Bulwark* with 42 Royal Marine Commando embarked at Karachi. An amphibious warfare squadron was based at Bahrain with Army units located at Sharjah, Bahrain, Aden, Kenya and Cyprus. Meanwhile, the RAF had two Hunter ground-attack squadrons (8 and 208) at Aden and Nairobi with light transport and communications at Bahrain and Aden (and heavier transports at Aden). Equipment reserves were stored at Bahrain and Kenya.

On 30 June the Emir of Kuwait appealed for help, the British Cabinet approved and

the Commander-in-Chief Middle East ordered landings to start as soon as possible after dawn the next day. On 1 July, a Beverley from 84 Squadron at Aden was the first military aircraft to touch down at Kuwait's civil airfield – a sand-covered strip in a sea of desert, 6 miles south of the town. On board were the groundcrew from 8 Squadron whose fighter ground-attack Hunter aircraft arrived from Khormaksar in Aden via Bahrain 10 minutes later in a sandstorm and less than 1,000m visibility. They were closely followed by another Hunter squadron (208) which arrived from Eastleigh, east of Nairobi via Bahrain.

On the same day, a pair of 37 Squadron Shackleton MR.2 aircraft moved up to Bahrain from Khormaksar to provide reconnaissance capabilities in the area.

The variety of RAF support for this operation can be gauged from the many different types of aircraft involved, including Comet, Britannia, Beverley, Valetta, Pembroke, Twin Pioneer and Hastings aircraft in the strategic and tactical transport role; fighter ground-attack attack and photographic reconnaissance Hunters; Canberra bombers; and photographic reconnaissance Canberra aircraft.

At the forward airstrip, in temperatures rising to almost 55°C, activity had been intense. The Hunters that had arrived fully armed with guns and rockets were back in the air as soon as refuelled, patrolling forward areas at the Iraq border where British and Kuwaiti troops were digging-in. An RAF signal squadron had set up an airfield control organisation, including a portable direction-finding station to help bring the aircraft 'home' in poor visibility. They also provided a telephone network, radio, teleprinter and Morse circuits with the Bahrain air head. An Air Movements team from Khormaksar worked non-stop with Army movements, unloading troops and equipment.

Two of the lessons learned during the Kuwait operation were the need for more and larger transport aircraft and for lighter and more modern air-portable equipment. The Beverley, with its spacious fuselage and ability to lift 19 tons from 300m roughly prepared strips, was a good workhorse for comparatively short journeys in the tactical area, but it could not carry the bulkier Army equipment that might have to go by air, and the RAF still had no bulk carrier capable of long-range strategic operations. Despite this, on the final tally when all the moves were complete, nearly 10,000 men and 850 tons of freight had been airlifted. The total cost of the operation was reported as being £900,000; at the time that seemed a small price to pay for the security of nearly half of Britain's oil supply.

Borneo

It was the ambition of Indonesia's President Soekarno to unite Malaysia, the Philippines and Indonesia within an Indonesian empire. Following the proposal in 1962 for a Malaysian Federation comprising Malaya, Singapore, Sarawak, Brunei and Sabah (North Borneo), Indonesia was determined to prevent the development. Brunei was considered a weak link in this proposal – the country was autocratically ruled by a sultan and was significantly richer than other proposed members due to its large oil revenues.

Soekarno encouraged the local North Kalimantan National Army (TNKU) to revolt. With 4,000 members, of whom 1,000 were under arms, the TNKU struck on 8 December 1962 and British military presence in Malaya was increased.

A detachment of three Shackleton MR.2s (WG555, WR964 and WR966) from 204 Squadron at RAF Ballykelly were sent to RAF Changi, arriving on 19 May 1964 where they undertook patrols to determine what Indonesian land forces were doing. Their stay lasted just 12 days and they returned to the UK. By August, the situation with what Soekarno referred to as a 'Confrontation' worsened significantly and Ballykelly was once again asked to support the aircraft of 205 Squadron based at Changi. Aircraft from 203 (WR965), 204 (WL739, WR964 and WR969) and 210 (WL748, WL751 and WL791) Squadrons were sent. The aircraft conducted reconnaissance operations over the Strait of Malacca under the code name Hawk Moth. In addition, the units operated a small force off the island of Pulau Labuan but, by October 1965, the last operational sortie had been flown and the aircraft returned to Ballykelly leaving only 205 Squadron operating from Changi.

On 9 August 1966, Singapore left the Federation and only weeks later, on 30 September,

ABOVE Avro Shackleton MR.2, WG530/G of 205 Squadron, over the coastline of Labuan, Borneo, during RAF Hawk Moth operations in the region, 1965. Note the white paint on the upper wings, an attempt to reflect the heat from the fuel tanks in the wings. *(Crown Copyright/Air Historical Branch image T-5269)*

BELOW The busy ramp at RAF Labuan during the Borneo confrontation in 1965 with a pair of Beverley C.1s (including XB289) from the Far East Air Force along with an unidentified Javelin and Canberra. The unidentified MR.2 in the foreground (coded H) was one of a number of Shackleton aircraft that conducted reconnaissance operations over the Strait of Malacca under the code name Hawk Moth. *(Crown Copyright/Air Historical Branch image PRB-1-29908)*

there was an attempted coup in Djakarta; the military counter-coup essentially ended Soekarno's rule and the prospect of peace seemed likely. Peace talks finally concluded on 11 August of the following year and the British forces were slowly withdrawn.

No 205 Squadron remained at RAF Changi until it was disbanded in October 1971, marking the end of Shackleton operations in the Far East.

British Guiana

In 1953, elections in British Guiana brought a result that many, including the governor, feared would lead to a communist government. The constitution was suspended and two companies of the 1st Battalion Royal Welsh Fusiliers were despatched in Royal Navy frigates.

In 1962, a general strike was called and further Army units were sent to Georgetown from Jamaica and the UK. A detachment of Shackleton MR.2 aircraft was sent from Jamaica in a show of force.

A small Royal Navy Wessex flight was based at Atkinson Field near Georgetown for support duties. In 1964, these were supplemented by three Whirlwind HC.10s along with Army Air Corp Auster AOP.9 aircraft from 24 Recce Flight. Around the same time, two Canberra PR.7 aircraft from 58 Squadron were detached from Piarco Airport, Trinidad, for survey work.

The colony became independent as Guyana in May 1966, under a coalition government. Most British servicemen left soon afterwards although a few did remain in order to train local forces.

Rhodesia and the 'Beira Patrols'

In 1963 the Central African Federation, comprising Northern and Southern Rhodesia and Nyasaland, broke up and the following year Northern Rhodesia and Nyasaland achieved independence as Zambia and Malawi respectively. Meanwhile, the white majority in Rhodesia had no intention of conceding to black majority rule and banned the two nationalist parties, the Zimbabwe African People's Union (ZAPU) and the Zimbabwe African National Union (ZANU). The principals of the parties were detained while the British government pressed for reform. The leader of the Rhodesian Front – Ian Smith – finally made a unilateral declaration of independence (UDI) on 11 November 1965.

The British response was to impose sanctions on imports and exports, a military solution committing British troops to fight whites being considered untenable by the government. Rhodesia was a landlocked country and was dependent to a large extent on the port facilities of its neighbour, Mozambique. The United Nations received reports that Rhodesia was acquiring copious supplies of oil, which were landed at the Mozambique port of Beira and transferred across the border. As a consequence, the Royal Navy brought in a number of ships to patrol the Indian Ocean's Mozambique Channel, between Mozambique and the island of Madagascar (then known as the Malagasy Republic).

BELOW Photographed at Sharjah during 1968 is Shackleton MR.3/3, WR983/E of 120 Squadron, which was detached to take part in the 'Beira Patrols' off the coastline of East Africa. *(Aviation Bookshop via Ray Deacon)*

Early in 1966, the British government had begun negotiating with the French for the use of Majunga airfield on Malagasy, but it was not until March that the first detachment of Shackleton MR.2s of 37 Squadron arrived from RAF Khormaksar. They were tasked with providing air reconnaissance of the area, especially suspicious shipping – details of which were relayed to the Royal Navy for them to intercept and board the offending vessels. These joint patrols were named 'Beira Patrols'. No 37 Squadron was followed by aircraft of 38 Squadron later in the year, and in February 1967 it was the turn of 42 Squadron. This arrangement continued until February 1972 when it was the responsibility of 204 and 210 Squadrons.

Each detachment, consisting of three Shackleton MR.2 aircraft with crews and ground personnel, lasted two months. Conditions were very poor with crews living in tents and fabricated metal huts known by their inhabitants as 'Camp Britannique'.

No 38 Squadron was disbanded in March 1967 with 37 Squadron following suit in September. A detachment of 42 Squadron MR.3s at St Mawgan stepped into the breach, but by April 1967, RAF Ballykelly had also been drawn into what became known as Operation Mizar and 210 Squadron provided a three-aircraft detachment. They were followed in turn by 204 Squadron from the same base, before 205 Squadron sent three aircraft from their base at Changi. The Shackleton involvement with Operation Mizar lasted until March 1972 when WL754 of 42 Squadron returned to St Mawgan.

While no country actually recognised Ian Smith's UDI, the Rhodesian High Court deemed the post-UDI government legal in 1968. The Smith administration initially professed continued loyalty to the Queen, but abandoned this in 1970 when it declared a republic in an unsuccessful attempt to win foreign recognition. The Rhodesian Bush War, a guerrilla conflict between the government and two rival communist-backed black nationalist groups (ZAPU and ZANU), began in earnest two years later, and after several attempts to end the war, Smith agreed to the Internal Settlement with non-militant nationalists in 1978. Under these terms the country was reconstituted under black rule as Zimbabwe Rhodesia in June 1979,

although this new order was rejected by the guerrillas and the international community. The Rhodesian Bush War continued until Zimbabwe Rhodesia revoked UDI as part of the Lancaster House Agreement signed in December 1979. Following a brief period of direct British rule, the UK granted independence to Zimbabwe in 1980.

South Arabia and the Radfan

South Arabia – from the Red Sea to Dhofar – was subject to British influence from 1839, with Aden becoming a Crown colony in 1937. In the hinterland there were constant inter-tribal conflicts, and after the success of air control in Iraq, the process was extended to the Aden Protectorate in 1927 with the posting of 8 Squadron. Problems in the region were dealt with by a combination of air policing and expeditions of ground forces, including the Aden Protectorate Levies (APL). Aden was a particularly important British base due to its

ABOVE AND BELOW
Two images of a Shackleton MR.2 (thought to be WR952/E) of 42 Squadron, Coastal Command, on operations against rebel tribesmen near Aden in 1957. No 42 Squadron, based at RAF St Eval, sent a detachment of four Shackleton MR.2s, which arrived at Khormaksar, Aden, on 7 January 1957. This started a rotation of aircraft through the base for operational duties that lasted for the next two years.
(Crown Copyright/Air Historical Branch image CMP-936)

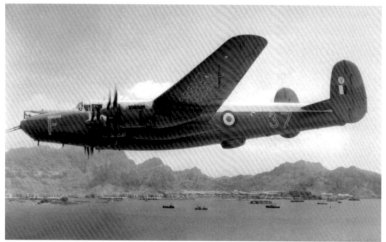

ABOVE After their rotation duties concluded, 42 Squadron Shackleton MR.2s continued to visit RAF Khormaksar for a number of years. MR.2/2, WL801/B, was photographed at Khormaksar during a month-long deployment in December 1962. During their time in Aden, they gained valuable front-line experience supporting 37 Squadron Shackletons on operations against insurgents in the rugged hinterland. *(Ray Deacon)*

LEFT No 37 Squadron became another Khormaksar temporary resident and like its predecessor (42 Squadron), worked closely with 8 Squadron on fighter-bomber operations. A 37 Squadron Shackleton MR.2, WR959/F, was photographed while flying over Aden Harbour on 16 March 1962 during formation rehearsals for an Air Ministry Film Unit. The photograph was taken from another Shackleton, WL744. No 37 Squadron remained at Khormaksar until the squadron was disbanded on 7 September 1967. *(Sandy McMillan via Ray Deacon)*

LEFT Another 37 Squadron Shackleton MR.2, WL744/B, at Khormaksar in the summer of 1962. This example had the Phase 1 modifications completed by 38 MU in January 1959 and Phase 2 modifications by January 1962. Owing to the excessive heat in Aden, special dispensation was given for the white tops of the aircraft fuselage to extend further downwards during the early period of Shackleton service, but this was withdrawn in 1963 when the markings reverted to the standard maritime pattern. *(Ray Deacon)*

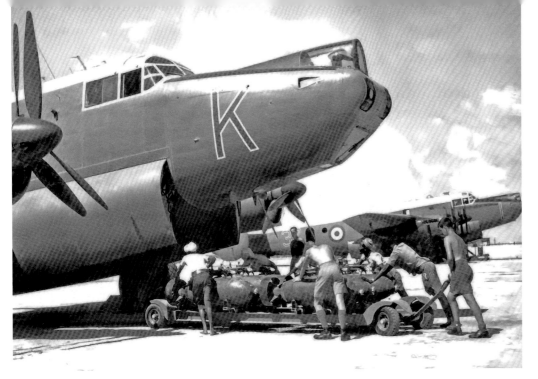

geographic location – on the route to India and between East Africa and the Gulf, both spheres of British influence.

Many of the major operations featured above captured the headlines, and although the Shackleton was involved in numerous activities in the Aden and Yemen Protectorates, they did provide the Shackleton with real 'shooting war' experience.

The Shackleton had replaced the Avro Lincoln of 1426 Flight in the colonial policing mission; aircraft would often be stationed in the Aden Protectorate and Oman to carry out various support missions, including convoy escorting, supply dropping, photo reconnaissance, communication relaying, and ground-attack missions. The Shackleton was also employed in several short-term bombing operations. Other roles included weather reconnaissance and transport duties; in the latter role each Shackleton could carry freight panniers in the bomb bay or up to 16 fully equipped soldiers.

No 42 Squadron, based at RAF St Eval, sent a detachment of four Shackleton MR.2s, which arrived at Khormaksar, Aden, on 7 January 1957. This started a rotation of aircraft through the base for operational duties that lasted for the next two years.

Many of these missions included air support and air cover for the ground forces, bombing, gunnery and acting as a radio link between the various forces in the area. An assortment of ordnance was carried on bombing missions,

ABOVE A 38 Squadron Shackleton MR.2, WL795/T, photographed at RAF Khormaksar in January 1964. Also visible to the left of the picture is Royal Navy Wessex HAS.1, XM844/02, probably from 815 NAS. *(Ray Deacon)*

BELOW No 210 Squadron from Ballykelly took its turn in the deployment to Khormaksar. Photographed in 1962 were two 210 Squadron examples, WL751/U to the left, with WL748/X at the rear. *(Ray Deacon)*

were also stationed at Khormaksar. Joint operations between the two squadrons became a regular occurrence, the Venoms and Hunters planning their strikes based upon aerial images obtained by the Shackletons. Little did they know that in the future, 8 Squadron would become the one and only squadron operating the Shackleton!

No 37 Squadron became another Khormaksar resident and like its predecessor, worked closely with 8 Squadron on fighter-bomber operations; 37 Squadron remained at Khormaksar until it was disbanded on 7 September 1967.

Belize

The former colony of British Honduras had long been threatened by neighbouring Guatemala, which had claimed rights to the territory. In 1948, it appeared that the Guatemalans might invade and two Royal Navy cruisers – HMS *Sheffield* and HMS *Devonshire* – were despatched with the 2nd Battalion the Gloucester Regiment. From that time, a company was deployed.

In October 1961 the country was devastated by Hurricane Hattie and additional troops were flown in from the UK via Kingston, Jamaica, in Operation Sky Help. Also participating were a pair of Shackleton MR.2s of 204 Squadron

ABOVE No 224 Squadron, usually resident at RAF Gibraltar, also took its turn in the rotation. Shackleton MR.2, WR951/W, was operated by 224 Squadron from October 1961 through to October 1966 when it was transferred to 204 Squadron. It was photographed while on deployment at RAF Khormaksar in June 1962. *(Ray Deacon)*

depending upon the individual targets. If intimidation was the purpose of a particular sortie, 20lb practice bombs were dropped and the Shackleton's vast bomb bay could accommodate a great number of these. A full load of 25 500lb, or 12 1,000lb bombs were used against more fortified positions, but the results of an attack were often disappointing unless a direct hit was scored.

Twelve-hour patrols were a feature of the Shackleton sorties and they were often the target of small-arms fire from the local tribesman. No 8 Squadron, equipped with a mix of Venom FB.1/FB.4s as well as Hunter FGA.9/FR.10s,

RIGHT Members of one of 37 Squadron's Shackleton MR.2 crews pose with their aircraft and a display of armaments at Khormaksar in August 1967, only weeks before the squadron disbanded. *(Alan Taylor via Ray Deacon)*

RIGHT Although details remain classified, it is known that SAAF Shackleton MR.3s flew some 200 hours on operational sorties during the Angolan civil war. This example, 1722, was preserved in flying condition by the SAAF Historic Flight at Ysterplaat AFB near Cape Town after the type had been withdrawn from service. Sadly, although the aircraft technically remains 'airworthy', it has since been restricted to the occasional ground run only. *(Peter R. March)*

along with two Valiant B(PR)K.1s from 543 Squadron for survey work.

In 1973, the country was renamed Belize. Further disputes with Guatemala continued and, in 1975, Guatemalan troops massed on the border. The British response was swift; three Puma HC.1s were deployed and the garrison increased to 1,000 men. Six Harrier GR.1As from 1 Squadron were based here for a short time before being returned to the UK. In June 1977, a similar situation occurred and six Harrier GR.3s made the Atlantic crossing. This time the British force remained with four Harrier GR.3s being resident on rotation as part of 1417 Flight based at Belize International Airport.

The main British force left Belize in 1994, three years after Guatemala recognised

Belizean independence, but the United Kingdom maintained a training presence via the British Army Training and Support Unit Belize (BATSUB) and 25 Flight AAC until 2011, when the last British forces left Ladyville Barracks – with the exception of seconded advisers.

SAAF in Angola

Although details remain classified, it is known that Shackletons flew some 200 hours on operational sorties during the Angolan civil war, some of which involved the protection of the hydroelectric works at Rucana.

LEFT A Shackleton MR.3, thought to be from 201 Squadron, photographed at RAF St Mawgan in April 1960. It demonstrates the ten-man flight crew, supporting groundcrew, along with weapons available for use and the ground equipment required to operate the aircraft. *(Crown Copyright/Air Historical Branch image T-1859)*

Chapter Four

An interim AEW solution

Following the disastrous consequences of the Japanese raid on Pearl Harbor on 7 December 1941, the US Navy took steps to ensure that such an event would never be repeated. Research undertaken by the General Electric Company (Westinghouse) eventually produced the AN/APS 20 radar, although its gestation period through numerous earlier iterations had been arduous.

OPPOSITE Shackleton AEW.2 WR963/63 of 8 Squadron, photographed along the Scottish coastline close to RAF Lossiemouth in 1990. This was taken shortly before the Shackleton AEW.2 aircraft were withdrawn from service in July 1991. *(Crown Copyright/RAF Lossiemouth/SAC Phil Ryan)*

However, some of the earliest General Electric (GE) technology did allow the US Navy *some* warning of the impending Japanese Kamikaze attacks – on their aircraft carriers in particular – towards the end of the Second World War.

Later, the GE AN/APS 20 radar system was positioned in a large ventral pod located between the main wheels of the Douglas AD-4W Skyraider, a three-seat airborne early warning (AEW) aircraft. Some 168 aircraft of this variant were ordered by the US Navy for fleet protection, each aircraft carrier operating its own complement of AEW Skyraider aircraft. This was followed by an improved version of the Skyraider, the AD-5W (later EA-1E), which was another three-seat AEW version with an APS-20 radar installed, of which 218 were built.

Through the Military Defence Assistance Programme (MDAP), the Royal Navy were supplied with 40 surplus former US Navy AD-4W AEW aircraft in 1951. All Skyraider AEW.1s – as they were designated in British military service – were operated by 849 Naval Air Squadron, which provided four-plane detachments for the British aircraft carriers. One flight aboard HMS *Bulwark* took part in the Suez crisis in 1956.

Fairey Gannet

The British military required their own, improved design and issued Specification GR17/45 for a shipboard, two-seat anti-submarine strike aircraft powered by an Armstrong Siddeley Double Mamba turboprop engine. Two prototypes of what would become the Gannet made their first flights on 19 September 1949 (VR546) and 6 July 1950 (VR557). They were followed by a third prototype (WE488), a three-seat version that first flew on 10 May 1951. Initially, 169 of the AS.1 version were ordered, along with 36 T.2 trainers. However, the possibility of an AEW version of the Gannet had been considered and an aerodynamic prototype (WJ440) made its maiden flight on 20 August 1958. This variant had an entirely new fuselage and tail unit, along with a crew of three – one pilot and two radar observers. A large radome to house the AN/APS 20 radar was mounted in the ventral position. This led to a production order for 44 aircraft designated Gannet AEW.3, with the first, XL442, making its maiden flight in December 1960.

Once in service, the Gannet AEW.3 soon replaced the Skyraider AEW.1 aircraft with 849 Naval Air Squadron. They were deployed on to the Royal Navy fixed-wing aircraft carriers – HMS *Victorious*, *Centaur*, *Ark Royal* and *Eagle* – and served the fleet well. However, the cost of operating such a large and expensive fleet of fixed-wing aircraft carriers was prohibitive while such a large asset was a relatively easy and effective cost-reduction target for the government of the day. One by one, the carriers were withdrawn until the then Labour government announced the total withdrawal of all fixed-wing aircraft carriers within the Royal Navy. The last left service when HMS *Ark Royal* docked at Devonport for the final time in November 1978. It effectively left the Royal Navy's remaining ships vulnerable, without any AEW capability.

Vulnerability in the Falklands conflict

Just how vulnerable became clearly apparent during the Falklands conflict, when Argentina invaded the Falkland Islands in 1982 and a British task force set sail to retake the islands. This Royal Navy task force had no airborne early warning capability, resulting in the well-publicised losses that were sustained from both Exocet missiles and conventional bombs. A hastily converted Sea King helicopter equipped with a Searchwater radar scanner in a retractable radome came too late to have any impact on the conflict.

Up to the announcement of the withdrawal of all Royal Navy fixed-wing aircraft carriers, the RAF had been happy to allow the Royal Navy to provide its own AEW facilities, as the RAF relied on the 'Chain Home' installations around the UK. However, the Royal Navy's decision to phase out all fixed-wing aircraft, leaving a serious gap in the fleet's defences, made many sit up and take notice. At the time, it was considered that the RAF could provide land-based AEW cover, although they had no aircraft capable of completing the task!

The political decision

However, back in 1967, it was agreed that a number of Shackleton MR.2 aircraft would be converted to the AEW role – purely as an 'interim solution' until a new AEW aircraft (expected to be the Nimrod AEW.3) entered service. The decision to refurbish some 15-year-old piston-engined aircraft was not considered too bizarre as it met with Treasury blessing – purely on a cost basis – while the NATO partnership viewed the decision with scepticism. NATO was, once again, assured that it was purely an 'interim solution', which would only be required 'for a few years'.

A check was made on all of the remaining Shackleton MR.2 airframes to identify the aircraft with the lowest flying hours. Twelve MR.2 Phase 3 aircraft were eventually identified and chosen for the conversion programme. Initially, all the chosen aircraft were flown to 5 MU at Kemble where each airframe was inspected and any repairs or maintenance issues completed. They were then placed in storage until called forward by Hawker Siddeley, who were to conduct the conversion programme at their Bitteswell facility. Fortunately, at around the same time, the Nimrod MR.1 was replacing Shackleton aircraft in the Maritime Reconnaissance role, with deliveries to MR squadrons commencing in October 1969, thereby permitting the availability of a wider choice of airframes.

No 8 Squadron was chosen to operate the Shackleton AEW.2. It had disbanded at Muharraq in December 1967 and was re-formed on 1 January 1972, initially at RAF Kinloss, as part of No 11 Group, Strike Command. At the time, the runway at Lossiemouth, the squadron's intended base, was undergoing repairs and the squadron officially moved there on 14 August 1973. Interestingly, 8 Squadron had previously operated the Hawker Hunter at Muharraq alongside 37 Squadron's Shackletons in Aden.

Two Shackleton MR.2 aircraft (WL787 *Mr McHenry* and WR967 *Zebedee*) were delivered to Kinloss in November 1971, in anticipation of the squadron's formation on 1 January 1972. These MR.2 aircraft enabled crew training to commence. As the whole AEW scene was new to the RAF, instructors from the Royal Navy's 849 Squadron were initially loaned to the Shackleton unit. With their experience on both the Skyraider and Gannet, the Navy had a surplus of trained operators and several of them became Shackleton crew members until the RAF had sufficient operators of their own checked out on the new avionics.

Conversion programme

The first aircraft in the AEW.2 programme was WL745, which initially began trials with the AN/APS 20F radar at Woodford in March 1970. Hawker Siddeley then completed the conversion process and the aircraft was delivered to the A&AEE at Boscombe Down for clearance trials in April 1972. The trials were completed by February 1973 and the aircraft was moved to Bitteswell to complete the conversion process before being the last AEW.2 aircraft delivered to 8 Squadron, on 19 September 1973, where the aircraft operated as *Sage*.

BELOW WL745 was the very first AEW.2 conversion, although it was the last aircraft to actually join 8 Squadron when it was delivered in September 1973. It remained with 8 Squadron until withdrawn from service in June 1981 – a victim of John Nott's defence cuts. *(Crown Copyright/Air Historical Branch image TN-1-6847-7)*

Radar installed

During the conversion process, the second-hand AN/APS 20F radars – previously operated within the Royal Navy Gannet AEW.3 fleet until the Treasury-induced decision to terminate all fixed-wing flying – found a second lease of life inside the Shackleton AEW.2. Two additional AC generators were fitted to the Nos 1 and 2 engines to power the unit and the additional electrical equipment carried. The radar system had an effective range of 200 miles. The scanner scope image was ground-stabilised from the aircraft Doppler, and was also north-stabilised so that north was always at the top of the display.

Another facet of the conversion programme was the increase in the number of the all-important 7in cathode ray radar operator positions – from two stations to three – while curtains were also installed around these three radar positions. The screens were difficult to operate unless extraneous light was kept to an absolute minimum. Most operators preferred to work in total darkness, especially when spending hour upon hour peering into the screens.

Even in the early 1970s, AEW operations were reliant on electronic countermeasures, so the RWR (radar warning receiver) Orange Harvest system (the white 'spark plug' shape on the roof) installed in Phase 3 Shackleton aircraft, was retained. Other electronic equipment systems introduced included the APX 7 IFF (identification friend or foe) along with both passive and active SIF (selective identification facility). Radio equipment comprised two PTR175 U/VHF, two R52 UHF and two Collins 618T single-side-band HF sets. The AEW.2 variant retained the huge bomb bay although this was usually fitted with smoke and flame flares along with Lindholme Gear.

The most noticeable external difference was the addition of the large ventral radome to house the radar, although the old 'dustbin' unit from the MR.2 was removed and the hole in the bottom of the fuselage was plugged. The 20mm cannon in the nose were also deleted.

Crew of nine

Initially, the AEW.2 had a crew of nine, consisting of pilot, co-pilot, radio navigator (operating the radios), 'navigating' navigator, air

ABOVE A close-up view of the AN/APS 20F radar scanner installation on WL757. The main lead is from the GPU (ground power unit), which supplies 28 volts DC. No 8 Squadron had modified Houchin GPUs that would produce a higher than standard output for those rather chilly and frosty winter mornings at Lossiemouth. A cold 36.7-litre supercharged Rolls-Royce Griffon with oil as thick as treacle would draw around 700-plus amps upon start. This GPU connection is purely for the main aircraft 28-volt system. There is another connection point (seen to the left of the main one), which supplies power to the radar equipment. On the ground, the radar would be selected to 'Dummy Load' to prevent accidental damage to personnel and ground radar equipment. This setting allowed the radar to function on the ground without transmitting. *(Keith Wilson)*

BELOW A wide-angle view of the cavernous bomb bay of the Shackleton AEW.2. In addition to the flares and Lindholme Gear, this space permitted the carriage of specially designed baggage panniers, enabling the groundcrew to carry essential spares and ground support items on deployments and exercises – ensuring they were generally self-sufficient. *(Keith Wilson)*

The 'Dodo' at Lossiemouth in February 1990. WR967 started life as a Shackleton MR.2 and was later delivered to 8 Squadron in January 1972 as a crew trainer with the name *Zebedee*. It suffered a serious accident on 7 September 1972 and later had the wings and tail removed to allow the fuselage to be converted into an AEW training simulator. It was allocated the maintenance serial 8398M, although it carried 'T83987' on the fuselage. *(Keith Wilson)*

engineer and four radar operators. The latter came in three groups: the basic AEW operator who was trained but required supervision; the more experienced operator who could work without supervision; and the tactical co-ordinator (TACO) who was the most experienced operator, also responsible for managing the other operators while in a tactical situation. After 1981, the radio navigator was replaced by an additional crew member. The 'off-duty' operator would provide frequent hot drinks and food from the small galley in the rear of the aircraft. The fifth mission crew member manned the radio when extra mission crew personnel replaced the second navigator in 1981.

AEW.2 deliveries commence

The first AEW.2 aircraft to join 8 Squadron was WL747 *Florence* (conversion No 3) which arrived at Kinloss on 11 April 1972. The following month WL756 *Mr Rusty* had been

LEFT The badge of the aptly named 'Dodo' – the extinct flightless bird – is carried on the starboard fuselage side of the AEW training simulator at RAF Lossiemouth. *(Keith Wilson)*

BELOW Shackleton AEW.2, WL747, was conversion No 3 and was delivered to 8 Squadron on 11 April 1972 where it operated as *Florence*. It remained with 8 Squadron until the Shackleton's final withdrawal in 1991 and was later sold to a private owner at Paphos International Airport where it resides today, albeit in a very poor condition. *(Crown Copyright/Air Historical Branch image TN-1-6698-118)*

ABOVE A line-up of six Shackleton AEW.2 aircraft of 8 Squadron, with WL756 nearest the camera, photographed at RAF Lossiemouth on 30 October 1973. With the defence cuts of 1981, forced upon them by Defence Secretary John Nott, the squadron's strength was reduced from 12 to just 6 airframes. Perhaps the worst part about it was that the squadron first heard of the announcement on the BBC's *6 O'clock News*! *(Crown Copyright/Air Historical Branch image TN-1-6847-9)*

BELOW LEFT Three of 8 Squadron's Shackleton AEW.2 aircraft on the ramp at RAF Lossiemouth on 16 December 1976. Nearest the camera is WL747, which has the *Magic Roundabout* character Florence painted just under the cockpit. *(Crown Copyright/Air Historical Branch image TN-1-7635-12)*

delivered and by the end of 1972 the squadron had received eight AEW.2 aircraft (see table on page 75). Further deliveries were made in 1973 until the final AEW.2 aircraft (of 12 ordered) was conversion No 1, which arrived at RAF Lossiemouth in September 1973.

Quick Reaction Alert (QRA)

Initially, and as far as the Royal Navy was concerned, 8 Squadron was considered to be the replacement for their own 849 Squadron Gannet AEW.3 aircraft. However, the yawning gap in the early warning facilities available to the United Kingdom Air Defence Region (UKADR) as a whole made it imperative that priority be given to it, above all other considerations. The aim of the UK Air Defence during that time was to maintain the integrity of the UKADR, which surrounded the UK and extended north towards Iceland, beyond the Faroe Islands and towards Norway. Any Soviet aircraft entering this area

LEFT Shackleton AEW.2, WL754, was the final of the 12 AEW.2 conversions delivered to 8 Squadron in November 1972, where it operated as *Paul*. It was photographed near to its RAF Lossiemouth base on 16 August 1977 after it had completed another North Atlantic patrol. *(Crown Copyright/Air Historical Branch image TN-1-7736-2)*

had to be intercepted by UK fighters, and fighter squadrons took turns to hold Northern and Southern QRA. Tanker forces also held alert to support the fighters and, if required, a Shackleton AEW.2 was also on call.

The 12 crews of 8 Squadron held a 2-hour alert, 24 hours a day, 7 days a week. Once a Soviet aircraft had been detected heading into the North Atlantic by Norwegian air defence sites, it was usual for the duty controller at RAF Strike Command to scramble the Shackleton to cover the Iceland–Faroe Islands gap. It could take up to 3 hours for the Shackleton to reach its barrier position and wait for the fighter scramble to take place. Often, the intruder would turn back before reaching UK airspace, but occasionally they would encroach.

The Shackleton operator would report the intruders via HF radio to one of the air defence stations – Saxa Vord, Polestar (on the Faroe Islands), Benbecula or Buchan. The controller would take control of the fighters and tanker aircraft before vectoring them towards the intruder. For 8 Squadron Shackleton crews, there was nothing more satisfying than to prosecute a gaggle of Soviet *Bear* aircraft with a pair of Lightning fighters!

During this time and particularly at the height of the Cold War, the relationship between the NATO powers and the USSR was, to the say the very least, poor. To put it bluntly, neither side trusted the other! The North Atlantic, with its approaches via the Norwegian and North Seas, witnessed a vast amount of naval activity by both sides, which attracted regular air force attention from each party. This location became 8 Squadron's main area of responsibility for many years, with its lengthy patrols over featureless seas, often in poor weather conditions, relieved only by the regular interception of Soviet *Bears* and *Bisons*.

ABOVE An interesting diversion for the author, during an otherwise busy operational sortie on board Shackleton AEW.2, WL757, on 15 February 1990, was the arrival of two Buccaneer S.2Bs from Lossiemouth-based 12 Squadron, for a brief photoshoot just off the coast of Aberdeen. Seen here is XX894 in 12 Squadron's special 75th anniversary colour scheme. The author was 'working' from the open observer's window on the starboard side of the fuselage, just above the number '8', while this image was shot by the navigator in a second Buccaneer aircraft. *(via Keith Wilson)*

BELOW A pair of RAF Wyton-based 100 Squadron Canberra TT.18s, WJ636/CX and WK118/CQ, photographed from the starboard fuselage observer's window of Shackleton AEW.2, WL757, at the end of their day's work. The pair of Canberra aircraft had provided the Shackleton radar operators with excellent one-on-one air-to-air combat training, known simply as 'knocking heads', the two aircraft taking their turn in being the aggressor or the defender. At the time of the photograph – 15 February 1990 – WJ636 was still painted in a special 100 Squadron 70th anniversary colour scheme, which had been added to the aircraft when the squadron celebrated the milestone back in February 1987. *(Keith Wilson)*

ABOVE The Tupolev Tu-20 *Bear-D* was a regular customer for 8 Squadron's Shackleton AEW.2 aircraft operating in the North Sea. A Russian *Bear* is escorted by Phantom FG.1, XV574/B, of 43 Squadron based at RAF Leuchars, during an interception made on 8 February 1973. *(Crown Copyright/ Air Historical Branch image TN-1-6716-35)*

ABOVE The unofficial 8 Squadron patch that was such a hit at airshows across the country. *(via Keith Wilson)*

Bear Hunter

The very large Tupolev Tu-95, with the NATO codename *Bear*, was powered by four Kuznetsev NK-12M turboprop engines of 12,000shp, each driving slow-turning contra-rotating AV-60N airscrews. When a Shackleton took up station alongside a *Bear*, there were 56 propeller blades thrashing the surrounding air. Such was the volume of sorties to find and intercept *Bears*, that the squadron, with its historic motto *Uspiam et Passium* (Everywhere Unbounded) adopted the unofficial designation 'Bear Hunter' and a special logo patch was produced which became a popular item at airshows.

RIGHT The sheathed Arabian dagger, correctly called a 'jambiya', features on the squadron badge (right) of 8 Squadron, along with the fighter pennant (left), as displayed on Shackleton AEW.2 WL757. *(Keith Wilson)*

The role of 8 Squadron

The task of an AEW aircraft is to detect, direct and report. The Shackleton crew operator would detect and report the position of radar contacts by voice radio to ground

LEFT The presentation of the roundel on an 8 Squadron Shackleton AEW.2 with the yellow, blue and red fighter bars on either side. The squadron number is presented in red with a white outline and is located on both fuselage sides, but is seen here just below the starboard rear observer's position. *(Keith Wilson)*

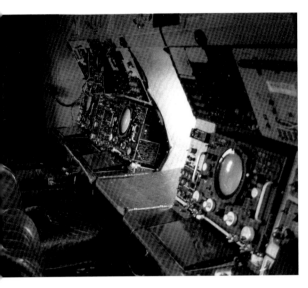

radar sites and ships – the procedure known as 'voice tell'. Once a hostile aircraft was detected, the controller would direct friendly fighter aircraft on to the target. Fighter Control was regularly practised with Phantom and Lightning fighters over the North Sea, working with the Sector Operations Centres (SOC) at Buchan, Boulmer and Neatishead.

For practice and training purposes, student controllers (both ground and airborne in a Shackleton) could call on the services of Canberra aircraft from 100 Squadron, who would act as both fighters and targets for training – a practice known as 'knocking heads' – witnessed at first hand by the author during a training sortie with an 8 Squadron Shackleton back in 1990.

As well as having the ability to detect aircraft, the Shackleton radar could also detect ships.

ABOVE LEFT The centre of operations in the AEW.2, showing the wing spar (middle-left of image) which divided the radar operators. During the AEW.2 conversion programme with Hawker Siddeley, the number of radar operator positions was increased from two to three stations. *(Keith Wilson)*

ABOVE To the right of the three radar stations was the position reserved for the tactical navigator. In this image, taken inside WL757 shortly before the aircraft departed for another operation, the aircraft is clearly placarded as 'Aircraft Armed'. *(Keith Wilson)*

Consequently, the crews had the capacity to carry out raids on hostile shipping by controlling attacking aircraft. Many hours were spent as 'Orange Forces' directing strikes by 12 and 208 Squadron Buccaneers against NATO navies on exercise around the British Coast.

A number of secondary tasks were given to 8 Squadron which provided variety to their

LEFT Four Shackleton AEW.2s of 8 Squadron, photographed in formation to mark a very special date for 8 Squadron – 8 August 1988 (8/8/88). The aircraft are **WL747, WL756, WL757 and WR963.** *(Crown Copyright/Air Historical Branch image DPR-550-5)*

flying. The Shackleton AEW.2 carried search-and-rescue equipment in its bomb bay, and Lindholme dinghy drop procedures were regularly practised.

Exercises and deployments

Joint RAF and NATO exercises were a regular feature of 8 Squadron's life, but at least one QRA aircraft had to be retained for possible interceptions. For the exercises, detachments were often deployed to St Mawgan to work in conjunction with NATO E-3As, although this often proved to be a technically one-sided affair. Despite their age, the AN/APS 20F radar coped well with modern high-speed aircraft, whose performance was far greater than those in the 1940s when Westinghouse developed the system. It was a credit to the skills of the radar operators on board the Shackleton aircraft who viewed the world outside through a 7in scope.

Other regular detachments made by the squadron included those to RAF Akrotiri where exercises were conducted with the MEAF and NATO fleets in the Mediterranean theatre. The other climatic extremes were encountered when the squadron sent small detachments to Iceland, for joint exercises in low-temperature environments. Here they exercised with the USAF forces in Keflavik, where the Shackleton would act as a target for the resident F-4E aircraft in Exercise Fan Angel.

Spoofing and communications jamming was employed by the Shackleton mission crew and, on one memorable occasion, the Shackleton TACO convinced two F-4Es that he was the *real* controller. On pulling them on to one of the squadron's own discrete radio frequencies, he split the aircraft into cloud, moved them around on fictitious contacts for 10 minutes, then finally talked the wingman into shooting down his leader!

ABOVE A fine study of two Shackleton AEW.2 aircraft from 8 Squadron (WR963 and WL757) photographed along the Scottish coastline close to their base at RAF Lossiemouth in 1990. *(Crown Copyright/SAC Phil Ryan)*

BELOW Another nice air-to-air portrait, this time of WR965, taken in 1987. Sadly, WR965 crashed in the Outer Hebrides on 30 April 1990, killing all on board. *(Geoff Lee/Planefocus image GL-SHACK-07)*

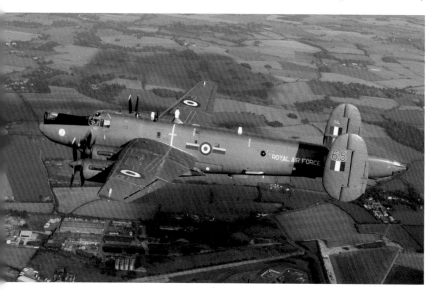

John Nott's defence cuts of 1981

In the spring of 1981, John Nott – the Defence Secretary for the then Conservative government – set another round of defence cuts into motion and 8 Squadron felt the full force of them. They were informed that the squadron was to lose half its aircraft, reducing their complement of twelve aircraft down to just six airworthy Shackleton AEW.2 aircraft. As a consequence, the six aircraft with the highest hours – WL741, WL745, WL754, WL795, WR960 and WR963 – were withdrawn from service. Two of them – WL741 and WL745 – were burnt, although some equipment, together with the engines, was salvaged for spares.

Thankfully, it ignited more determination than ever within 8 Squadron to keep their handful of aircraft at constant readiness. To their immense credit, this was accomplished throughout the next ten years.

Cancellation of the Nimrod AEW.3

As mentioned earlier, the British government had decided to go it alone in developing a suitable AEW aircraft, having declined an earlier offer from Boeing for E-3 Sentry aircraft around the same time that NATO had continued to procrastinate over their acquisition of the E-3.

A de Havilland Comet 4C, which had once been flying with British Overseas Airways Corporation (BOAC) as G-APDS, now carried the military markings XW626. It was stripped and refurbished by BAe at Woodford in the early 1970s to become a flying laboratory for the development of the AEW trials. The Royal Radar Establishment (RRE) at Pershore was technically responsible for the design and installation of the crew stations, system racks and cabling, along with what would become the largest nose radome ever constructed at that time in the West. Interestingly, although the Nimrod AEW.3 was designed with a radome for the tail, it was not fitted to XW626. The aircraft made its first flight in June 1977 and trials continued through to 1980. At the successful conclusion of the programme, XW626 was transferred to the RAE at Bedford for further trials work.

8 SQUADRON'S MAGIC ROUNDABOUT

8 Squadron's Shackleton AEW.2 aircraft were all named after characters from *The Magic Roundabout*, a BBC children's television programme that earned a cult following with many adults in the UK in the 1960s.

WL741	AEW.2	*PC Knapweed*
WL745	AEW.2	*Sage*
WL747	AEW.2	*Florence*
WL754	AEW.2	*Paul*
WL756	AEW.2	*Mr Rusty*
WL757	AEW.2	*Brian*
WL787	MR.2	*Dylan*
WL790	AEW.2	*Mr McHenry* (later renamed *Zebedee*)
WL793	AEW.2	*Ermintrude*
WL795	AEW.2	*Rosalie*
WR960	AEW.2	*Dougal*
WR963	AEW.2	*Parsley* (later renamed *Ermintrude II*)
WR965	AEW.2	*Dill* (later renamed *Rosalie*)
WR967	MR.2	*Zebedee*

BELOW The rather hasty introduction of the Shackleton AEW.2 had always been considered an 'interim solution'. However, the development and introduction of its proposed replacement – the Nimrod AEW.3 – was fraught with difficulties, especially with the GEC Avionics suite. It was eventually cancelled in January 1987 when orders were placed with Boeing for seven E-3D Sentry AEW.1 aircraft. This image, taken during a more optimistic period in the Nimrod AEW.3's development programme, shows XZ285 of the Nimrod AEW Joint Trials Unit in formation with WR960/60, a Shackleton AEW.2 of 8 Squadron. (*Crown Copyright/Air Historical Branch image TN-1-9188-8*)

In April 1977, the MoD ordered 11 Nimrod AEW.3 aircraft, although only two would be new-build aircraft, the remaining nine scheduled for conversion from the existing RAF fleet of Nimrod MR.1s. The first new-build aircraft was XZ286, which made its maiden flight from Woodford on 16 July 1980. Development work proceeded quickly with the second aircraft (XZ287) making its first flight in January 1981. Meanwhile, BAe had commenced the conversion programme and the first completed aircraft (XZ281) flew in January 1980. At this stage, things looked good for the Nimrod AEW programme to replace the ageing Shackleton.

Shortly afterwards, in January 1982, NATO received its first E-3A AWACS aircraft. In the meantime, back in the UK, the radar system – being developed by GEC Marconi – was suffering with 'slippage', owing mainly to radar clutter. The Argus radar system was a little unusual in as much as it featured two mechanical scanners, which provided 360° coverage of both air and surface targets. The scanners in the Nimrod AEW.3 were located in radomes at either end of the aircraft – significantly different from the single, large radome dish on the Boeing AWACS design. Software difficulties were suffered with a lack of co-ordination of the two scanners and much of the problem was laid at the door of the GEC 4080M onboard computer. It was estimated that a 300% increase in computing capacity would be required to find a solution; even then, the 300% increase would not provide the radar with any potential for improvement in future development.

By the end of 1984, with serious radar and software delays still apparent, it was calculated that the whole project had so far cost £816 million. A new fixed-term contract was put in place that provided for a solution by 'some time in 1987'. By the end of 1986, three development aircraft were involved in trials with the Joint Trials Unit (JTU) at RAF Waddington.

Despite pressure from political, military and business quarters, the JTU was able to clearly demonstrate that the system did not work and in January 1987 an emergency meeting was held in the House of Commons. The outcome was a majority of 169 for the government who announced the cancellation of the whole Nimrod AEW.3 programme. All of the Nimrod AEW.3 airframes were scrapped as it was considered too expensive to return them to the original MR.1 configuration. In all, over £900 million had been spent on the programme.

In 1988, the British government accepted a proposal from Boeing for the supply of seven E-3D Sentry AEW.1 (AWACS) aircraft at a cost of £860 million.

Disaster

Meanwhile, despite all of the political and industrial ramifications, the remaining six Shackleton AEW.2 aircraft continued to serve at RAF Lossiemouth. Sadly, that number was reduced when, on 30 April 1990, WR965 *Rosalie* crashed into a hillside near Northton, on South Harris in the Outer Hebrides, with the loss of all ten crew members including 8 Squadron's Commanding Officer, Wing Commander S.J. Roncoroni.

The crash had a sobering effect on the whole squadron and the new Commanding Officer – Wing Commander Chris Booth – made it his duty to reinstall the unit's pride in what they were doing.

Arrival of the Sentry AEW.1

The prototype Sentry AEW.1 made its maiden flight on 16 June 1990. The first aircraft to arrive at Waddington was ZH102 (not the prototype), which arrived early in November 1990. It was fitted out at Waddington and formally handed over to the RAF at Waddington in March 1991. The STS had been formed at RAF Waddington on 1 July 1990. The aircraft was fitted out with radars, computer consoles

BELOW When the Boeing E-3D Sentry aircraft started to arrive at RAF Waddington in June 1990, the end for the Shackleton AEW.2 really was in sight. By March of the following year, 8 Squadron's migration to the Sentry was completed. Here, the first Sentry AEW.1 was photographed in formation with Shackleton AEW.2, WL757, during April 1991. *(Geoff Lee/ Planefocus image GL-9101039)*

and communications systems by a team from Boeing and BAe technicians. The official handover to the RAF took place at Waddington in March 1991 and the remaining aircraft arrived at two-monthly intervals.

In the meantime, elements of 8 Squadron had moved down from Lossiemouth to Waddington ahead of the official handover on 30 March 1991. The Shackleton AEW.2 was no longer required.

Farewell Shackleton

The retirement of the Shackleton AEW.2 had been forecast at regular intervals, but it did eventually materialise in 1991, not at its home base of RAF Lossiemouth, but at its new home at RAF Waddington.

It had been 30 years since the Shackleton was described as 'worn out and obsolete' by the media and 19 years after being established as an 'interim solution' to the AEW problem.

The remaining aircraft were officially withdrawn from service on 1 July 1991. WL756 was flown to RAF St Athan for crash rescue training and had been completely destroyed by fire in 1998. Thankfully, the cockpit section was saved by Shackleton aficionado Druid Petrie and has since been sold to a collector in Holland.

WL747 and WL757 were flown down to Paphos, Cyprus, where they remain today, although both are in a very poor condition due primarily to the close proximity of the sea.

WL790 was sold to Air Atlantique at Coventry and flown to the USA in 1994, where it continued to entertain crowds at air displays for a further 14 years. The aircraft is now a ground exhibit at the Pima Air & Space Museum in Tucson, Arizona, and sadly, is not expected to fly again.

The 6,848-day 'interim solution' had finally come to an end.

ABOVE Later, during the same photoshoot, the Sentry AEW.1 and Shackleton AEW.2 were joined by a pair of 3 Squadron Tornado F.3 aircraft (ZG730/CC and ZE254/CA) from RAF Coningsby. (Geoff Lee/Planefocus image GL-9101056)

8 SQUADRON'S SHACKLETONS

Conversion	Serial	Delivery date	Date of withdrawal
Conversion 1:	WL745	17 September 1973	June 1981
Conversion 2:	WL756	5 May 1972	1 July 1991
Conversion 3:	WL747	11 April 1972	1 July 1991
Conversion 4:	WR960	8 June 1972	November 1982
Conversion 5:	WR963	19 July 1972	1 July 1991
Conversion 6:	WL757	29 August 1972	1 July 1991
Conversion 7:	WL790	23 September 1972	1 July 1991
Conversion 8:	WL795	20 October 1972	24 November 1981
Conversion 9:	WL741	4 April 1973	29 May 1981
Conversion 10:	WL793	23 February 1973	1981 (to 8675M)
Conversion 11:	WR965	31 January 1973	Crashed 30 April 1990
Conversion 12:	WL754	29 November 1972	22 January 1981
MR.2 crew training	WG556	5 May 1977	1980 (to 8651M)
MR.2 crew training	WL738	8 March 1974	14 October 1977 (to 8567M)
MR.2 crew training	WL787	1 January 1972	3 January 1974
MR.2 crew training	WL801	15 August 1974	June 1979
MR.2 crew training	WR967	1 January 1972	Accident 7 September 1972 Later converted to AEW training simulator as 8398M

Chapter Five

Anatomy of the Shackleton

The basic structure of all variants of the Shackleton was similar and followed Avro's traditional and well-proven practices developed during the manufacture of the Manchester, Lancaster and Lincoln bombers. It was this simplicity of modular construction that permitted a relatively easy assembly, but also provided a significant benefit in the event of an aircraft requiring repair after an accident.

OPPOSITE The Shackleton Preservation Trust's Shackleton MR.2/AEW.2 at Coventry during the first engine run of 2015, which took place on 21 March. *(Keith Wilson)*

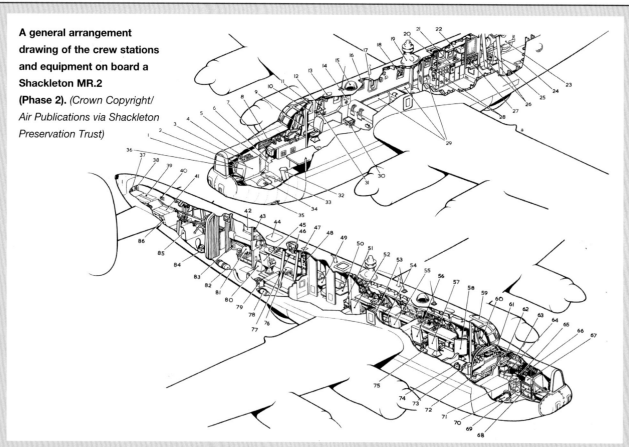

A general arrangement drawing of the crew stations and equipment on board a Shackleton MR.2 (Phase 2). *(Crown Copyright/ Air Publications via Shackleton Preservation Trust)*

1 Inverter control panel
2 Electrics crate
3 Fire extinguisher stowage
4 Crash helmet stowage
5 Windscreen de-icing tank
6 Emergency battery stowage
7 Autopilot amplifier and junction box
8 Ammunition container
9 Second pilot's seat
10 Signal cartridge stowage
11 Signal pistol (in stowage)
12 Drift recorder
13 Loop aerial
14 First-aid kit and axe stowage
15 Astrodome
16 Intercom controller
17 Ditching exit
18 Window (life raft forward release point)
19 ARI 18144 aerial plinth
20 Mid-upper escape hatch
21 Galley
22 Drinking water tank

23 Main door
24 Flame float launching tube door
25 Starboard observer's seat
26 Scanner jacks
27 Flare dischargers
28 Fire extinguisher and axe stowage
29 Dinghy and parachute stowages
30 Front spar step
31 Engineer's crash harness stowage
32 Flare chute
33 Air bomber's cushion
34 Parachute exit
35 Ram air intake – No 1 heater
36 Gunner's seat
37 Astro-compass
38 Curtain
39 Rear observer's cushions
40 Camera cupola electro-pneumatic control valves

41 Camera access doors
44 Portable water pump (stowed)
45 'Blue Silk' inverter compartment
46 Warning horn
47 Rear upper escape hatch
48 Port observer's seat
49 Photoflash dischargers
50 Pyrotechnic stowage
51 Bunks
52 Folding table
53 Seat
54 ASV operator's station
55 Secondary sonic's station
56 Subsidiary sonic's station
57 Master sonic's station
58 Routine attack navigator's station
59 Tactical navigator's station
60 Signaller's station
61 First pilot's escape hatch
62 First pilot's seat
63 Control column

64 Rudder pedals
65 Ammunition container
66 Flare chute control panel
67 Autolycus equipment
68 Pyrotechnic stowage
69 Bombsight computer
70 Pneumatic crate
71 Windscreen water tank
72 No 1 heater
73 Radio crate
74 Air-conditioning ducts
75 Engine controls pedestal
76 Trim controls
77 No 4 heater
78 MS9 life raft
79 Flame float ejector
80 'Blue Silk' equipment
81 Emergency pack
82 First-aid kit
83 Periscope (stowed)
84 Elsan chemical closet
85 Relief tank
86 Camera cupolas

RIGHT Shackleton MR.3/3, XF708, photographed outside on the ramp at Duxford before the aircraft was taken inside the Restoration Hangar to undergo a major overhaul. Prominent in this view is the pair of 20mm cannon in the nose.
(Keith Wilson)

Commencing with the Shackleton MR.1, the fuselage was manufactured in five distinct sections: nose; front centre section; intermediate centre section (which included the inner wing and inboard engine mountings); rear centre section and rear fuselage (including tailplane attachment points). Each section was of a light-alloy monocoque structure formed from longitudinal angle-section stringers and oval-shaped transverse formers mounted on a floor structure which incorporated two heavy longerons, cross beams and intercostals.

All five sections were bolted together utilising specially strengthened end formers providing the flange joints. The method of assembly also provided significant benefits when the aircraft had to be dismantled for any reason, such as accidental damage, which could be rectified thanks to the clever but simple modular construction and assembly process.

The nose section, forward of the cockpit, was heavily glazed to provide a good observation position for the bomb-aimer/observer seated on a wide, padded bench. The second crew position had originally been intended for a gunner operating the front gun barbettes (see Chapter 1, page 16) originally fitted on the prototype airframes but later deleted from the production orders.

The top of the front centre section was cut away to accommodate light-alloy cockpit framing, which incorporated escape hatches in the roof. The windscreen was of the dry air sandwich-type, the remainder of the glazed panels being of heavy-duty Perspex.

LEFT The nose of Shackleton WR963 at Coventry. This aircraft has had its AN/APS 20 radar and radome removed from underneath the nose and replaced with a cover. The ASV-21 radome has also been replaced in the lower rear fuselage, returning the aircraft's external appearance to its original MR.2 configuration. The 20mm cannon were removed from the MR.2 Shackletons on conversion to AEW.2 variants, although WR963 has since had 'dummy' cannon fitted.
(Keith Wilson)

Avro Shackleton MR.3.

(Mike Badrocke)

1. Starboard rudder construction
2. Rudder tabs
3. Rudder horn balance
4. Fin construction
5. Starboard tailplane construction
6. Reconnaissance camera
7. Strike camera
8. Rear observer's couch
9. Tailcone observation window
10. Astro compass
11. Fuselage tailcone
12. Tailplane centre section joint
13. Elevator tabs
14. Port elevator
15. Port rudder
16. Rudder cable control
17. Port tailfin
18. HF aerial cable
19. Tailplane leading edge de-icing
20. Port tailplane
21. Rear cabin curtained bulkhead
22. Tailcone access decking
23. Waste water tank
24. Toilet
25. Emergency equipment pack
26. Bomb bay periscope
27. Coat rail
28. Cabin wall trim panelling
29. First-aid kit
30. Control cable ducting
31. 'Blue Silk' radar equipment
32. Rear entry door
33. Emergency equipment stowage
34. Flame float ejector
35. Observation windows, port and starboard
36. Rear cabin escape hatch
37. Observers' seats, port and starboard
38. Rear cabin heater
39. Heat exchanger air intake
40. Ventral 'dustbin' radome, extended
41. Dinghy stowage
42. Bomb door hydraulic jack
43. Flare discharge chute
44. Flare dischargers
45. Flare stowage rack
46. Ventral radome extending hydraulic jacks (two)
47. Photoflash discharger
48. Fresh water tank
49. Pyrotechnic stowage rack
50. Crew rest bunks
51. Galley unit
52. Crew rest area
53. Cabin roof escape hatch
54. Folding table
55. Aft facing seat
56. Operational cabin rear bulkhead
57. Dinghy and parachute stowage
58. Wing trailing edge rib construction
59. Life raft stowage
60. Inboard nacelle tail fairing
61. Flap shroud ribs
62. Split trailing edge flap
63. Flap operating shaft
64. Rear spar
65. Fuel jettison pipe
66. Four-segment aileron
67. Aileron hinge link
68. Aileron tabs
69. Starboard fixed tip-tank
70. Fuel filler cap
71. Tip-tank rib construction
72. Pressure relief valve
73. Starboard navigation light
74. Wing rib construction
75. Front spar
76. Leading edge nose ribs
77. Leading edge de-icing strips
78. Wingtip panel joint rib
79. Port outer wing fuel tanks; total internal fuel capacity 3,916 imp gal (17,802 L)
80. Rolls-Royce (Bristol-Siddeley) Viper 203 boost engine (Phase 3 aircraft only)
81. Water methanol tank
82. Boost engine air intake, open
83. Hydraulic reservoir
84. Engine bay fireproof bulkhead
85. Engine bearer struts
86. Exhaust muffler
87. Radiator air outlet duct
88. Engine cowling panels
89. Oil and coolant radiators
90. Spinner
91. Forward retracting twin mainwheels
92. Mainwheel doors
93. Main undercarriage leg strut
94. Intermediate fuel tank
95. ASV 21 search radar scanner
96. Outer wing panel joint rib
97. Main undercarriage wheel bay
98. Engine mounting ribs
99. Fuel tank bay access panel
100. Inboard fuel tank
101. Wing/fuselage attachment main frames
102. Ditching hatch port and starboard
103. Secondary sonic operator's seat
104. Rear spar centre section carry-through housing
105. ASV operator's seat
106. ASV instrument displays
107. Overhead handrail
108. ARI 18144 ECM aerial
109. Sonics equipment racks
110. Master sonic's station
111. First-aid kit
112. Front spar centre-section carry-through housing
113. Attack navigator's seat
114. Plotting table
115. Instrument displays
116. D/F loop aerial
117. Astrodome observation hatch
118. VHF aerials

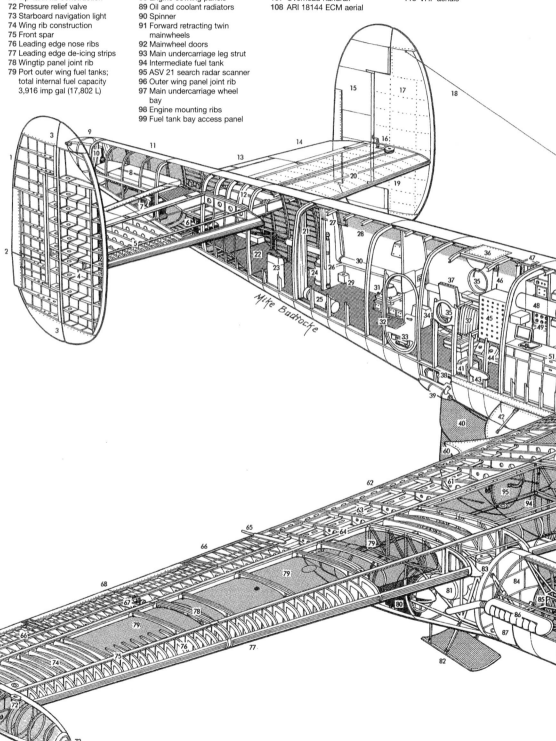

119 Port wing inboard fuel tank
120 Port split trailing-edge flap
121 Shackleton AEW Mk 2 nose profile
122 Radome for early warning radar
123 AN/APS 20 radar scanner
124 Fuel jettison pipe
125 Port four-segment aileron
126 Aileron tabs
127 Port tip-tank
128 Fuel filler cap
129 Port navigation light
130 Wing leading edge de-icing
131 Port outer wing fuel tanks
132 Outboard engine nacelle
133 Detachable cowling panels

134 Oil and coolant radiator intakes
135 De Havilland six-bladed constant-speed contra-rotating propeller
136 Frameless cockpit canopy cover
137 Cockpit roof escape hatch
138 Port side signaller's station
139 Folding sun blind
140 Tactical navigator's station
141 Flight engineer's station
142 Cockpit bulkhead
143 Co-pilot's seat
144 Engine throttle and propeller control levers
145 Control column

146 Instrument panel
147 Pilot's seat
148 Instrument panel shroud
149 Windscreen panels
150 Windscreen wipers
151 Rudder pedals
152 Cockpit floor level
153 Ammunition tanks, port and starboard
154 Pyrotechnic stores racks
155 Air gunner's seat
156 Gunsight
157 Gun control panel
158 Optically flat gunner's windscreen
159 20mm cannon (Phase 2 aircraft only, deleted on Phase 3)
160 Gun swivelling and elevating mechanism
161 Ammunition feed chutes
162 Ballast weights
163 Ventral air bomber's observation window
164 Bombsight
165 Radio homing aerial

166 Nose undercarriage leg strut
167 Twin nosewheels
168 Nosewheel doors
169 Hydraulic retraction strut
170 Air bomber's prone position couch
171 Forward radio rack
172 Starboard side electrical equipment rack
173 Windscreen spray and humidity system water tank
174 Underfloor control runs
175 Forward cabin heater
176 Ventral entry hatch
177 Boarding ladder
178 Bomb bay door
179 Bomb door hydraulic jack
180 Door frame construction
181 Bomb door hinge beam
182 Cabin floor beam construction

183 Cable duct
184 Drift recorder
185 Bomb bay overload fuel tank, 400 imp gal (1,818 L)
186 Rolls-Royce Griffon 57A V12 engine (Phase 2 aircraft; Griffon 58 in Phase 3 aircraft)
187 Exhaust collector
188 Oil and coolant radiators
189 Engine oil tank
190 Propeller reduction gearbox
191 Propeller hub pitch change mechanism
192 Spinner
193 1,000lb (454kg) HE bomb
194 Mk 30 torpedo
195 Torpedo parachute container
196 Mk 11 depth charge
197 Smoke marker
198 Parachute flare

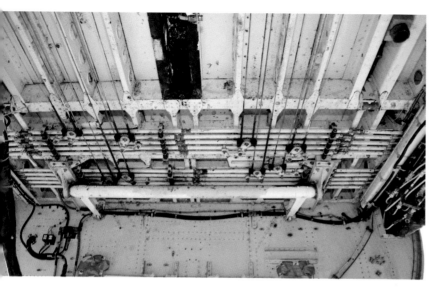

was lit by an ultraviolet light which illuminated fluorescent paint on the instrument dials. The remaining instruments were lit by red lighting. Dual flying controls with pendulum-type rudder pedals and handwheel control columns were provided. Here, tabular push-rods were used for control operation except for the ailerons, which were moved by a series of chains, tie rods and cables. The aircraft also featured an electrically operated Mk 9 autopilot, with the controller built into the first pilot's right-hand arm rest.

Bomb bay

The massive bomb bay was 33ft 6in long, 7ft 6in wide and 4ft deep. It was enclosed by two hydraulically operated doors. Detachable inserts in the door were provided to allow a Mk 3 airborne lifeboat (see Chapter 1 page 19) to be suspended from the centre bomb station. While this lifeboat facility was designed and tested, it was never used operationally as the Lindholme Gear – which could be installed within the bomb bay, thereby not affecting the aerodynamics of the aircraft – was the preferred option.

ABOVE The forward edge of the bomb bay on Shackleton WR963 features a number of cable and actuator rods. The main white bar just below the centre is the connecting rod between the two pilots' flying control yokes. *(Keith Wilson)*

RIGHT The cavernous bomb bay in Shackleton MR.3/3, WR974/K, at Charlwood. The clever design, utilising up to five bays – each bay having a variety of stations – permitted a flexible load-carrying capacity. In extreme cases, these *could* have included the carriage of a single 'Grand Slam' bomb! *(Keith Wilson)*

FAR RIGHT A single, electrically actuated bomb rack, positioned on the starboard side of bay 3, inside the bomb bay of WR963 at Coventry. *(Keith Wilson)*

Flying controls

Full flying controls and a standard blind flying panel were provided for each of two pilot positions. The engine instruments were located in the bottom centre of the panel, between the two seats and clearly visible from both. This panel was located directly above the access way into the nose section. The blind-flying panel

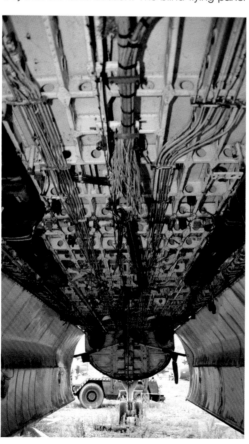

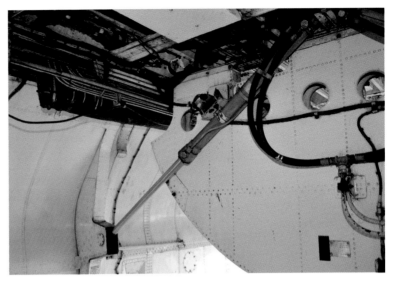

Wing

The cantilever mainplane was manufactured in five sections: centre portion (integral with the fuselage); port and starboard intermediate sections; port and starboard outer sections, each carried on a front and rear spar. The aerofoil shape at the wing root was NACA 23018 with a 4° dihedral applied to the intermediate and outer sections, which were heavily tapered. The wingtips were detachable. The light-alloy spars were built up from extruded U-section, machined from extruded bar and a heavy sheet web. Each section was covered with stressed light-alloy skin and joined by forged steel shackles and high-tensile steel bolts.

TKS porous de-icing strips were fitted in the leading edges of the intermediate and outer planes and fed with fluid via a metering pump from a 23-gallon tank mounted in the port wheel bay.

BELOW Shackleton MR.3, XF711, on the production line at Avro's Manchester production facility in early 1959. The aircraft was delivered to 201 Squadron as 'L' in May 1959. Visible in this image are the flaps and the starboard wingtip showing the service connections and starboard aileron actuators. *(Avro Heritage Collection image 3980l)*

RIGHT With the outer wing removed from Shackleton MR.3/3, XF708, during restoration work at Duxford, it exposes the inner fuel tank installation. Also visible in this view are the electrical connections at the top left of the image. *(Keith Wilson)*

Ailerons and flaps

The ailerons and flaps were also constructed from a light alloy, the ailerons being manufactured around a flanged pressed-sheet spar with mass balance weights bolted to the leading edge. Each aileron was in two sections, coupled by a torque tube to allow flexing, and was fitted with a trim tab and balance tab.

The split trailing edge flaps were manufactured in a similar manner to the ailerons. They were hinged to the rear spar and were in two sections: an inboard section from the fuselage to the aft fairing of the inner engine nacelle, and an outboard section from the centre/intermediate wing section joint out to the ailerons. They were operated by push-pull tubes and actuating rods, which allowed the spanwise action of a single hydraulically operated jack mounted transversely in the fuselage to provide an up and down movement at the flap.

Fuel tanks

Four fuel tanks were provided in each wing. The first was a crash-proof rigid tank of 497-gallon capacity, while the three remaining tanks were of the self-sealing flexible variety containing 541, 297 and 311 gallons respectively. Overall, it afforded a normal fuel capacity of 3,292 gallons.

Undercarriage

The main undercarriage beams were bolted in the front spar, each assembly consisting of two Dowty oleo pneumatic shock-absorber legs connected at their lower end by the wheel axle and at the upper end by a braced cross strut. Each axle carried a 64in-diameter Dunlop wheel and twin pneumatic brake units operated at 180lb/sq in. The wheel bay doors were connected to the shock-absorber legs so they

FAR LEFT One of the major differences between the MR.1/MR.2/AEW.2 variants of the Shackleton and the later MR.3 is the configuration of the undercarriage. All of the earlier aircraft had a retractable tailwheel configuration while the later MR.3 had a retractable nosewheel system. Shown here is the port main undercarriage on Shackleton MR.2/AEW.2, WR963, at Coventry. Each axle carried a 64in-diameter Dunlop wheel and twin pneumatic brake units operated at 180lb/sq in. *(Keith Wilson)*

LEFT The same undercarriage leg, featuring a single mainwheel; this time it is viewed from inside the undercarriage bay showing a number of the services required to operate the system. The two red bars are jury struts fitted after landing and used to secure the undercarriage while on the ground, which prevented an unplanned undercarriage collapse. The green bottle in the top left of the picture is one of the fire extinguisher bottles. *(Keith Wilson)*

opened and closed mechanically on selection of the undercarriage operating mechanism. In an emergency, the undercarriage could also be opened by compressed air if the electro-hydraulic system failed. The same system was capable of opening the bomb bay doors and wing flaps.

The main undercarriage assemblies retracted into the inner engine nacelles and when in the 'up' position both were covered by flush-fitting doors.

On the MR.1 and MR.1A, the fully castoring non-retractable tailwheel was supported by a longitudinal beam through a long shock-absorbing strut, which included a self-centring device. A fluid dashpot was fitted to prevent 'shimmy' of the single 32in-diameter tailwheel. This was later modified on the MR.2/AEW.2 to feature a twin-wheeled fully retractable undercarriage.

LEFT The large pneumatic brake system on the port main undercarriage leg of Shackleton MR.2/AEW.2, WR963, at Coventry. *(Keith Wilson)*

LEFT The neat retractable tailwheel undercarriage on WR963, viewed looking towards the tail. *(Keith Wilson)*

BELOW The main undercarriage leg of the Shackleton MR.3 variant featured twin mainwheels on either leg. It can be seen here on XF707 at Duxford looking towards the nose of the aircraft. These retracted forward into the wheel wells. *(Keith Wilson)*

LEFT The sturdy, twin-wheel nose leg on MR.3/3, XF708, which retracted rearwards into the bay. The early days of MR.3 operations were marred by a series of nosewheel failures on landing. This image was taken while standing in the undercarriage bay looking towards the nose of the aircraft. *(Keith Wilson)*

ABOVE The twin-finned tailplane showing the large rudders with two rudder trim tabs on each to good effect on WR963. All variants featured two rudder trim tabs on each end plate. *(Keith Wilson)*

ABOVE RIGHT The rear observer's position on the nosewheel MR.3/3 is subtly different from that on the tailwheel MR.2. The rear fuselage also features retractable doors for both the strike and reconnaissance camera positions, which can be seen looking directly up on MR.3/3, XF708. *(Keith Wilson)*

BELOW The mighty Rolls-Royce Griffon engine. This is a model RG-30 SMS and was photographed after overhaul at Retro Track and Air Limited. This version was originally used in the Shackleton but has since been modified for use in the Battle of Britain Memorial Flight Spitfire PR.XIX aircraft. *(Keith Wilson)*

Tailplane

The two-spar tailplane was built in two sections, bolted together on the centreline of the fuselage, to which the complete assembly was attached using high-tensile bolts. Pressed ribs and skinning from a light alloy completed the assembly. The elevators were manufactured in a similar manner with a single 'nose' spar to which lugs were bolted, joined with forked attachments on the rear spar of the tailplane to form the hinges. The end plate fins were built up on two posts, which had web plates riveted to angle-section booms and light ribs to form an aerofoil shape. The front and rear posts were bolted to the ends of the tailplane spars, horn-balanced rudders being hinged to the rear fin posts and fitted with two large trim tabs operated by a duplicated handwheel in the cockpit. TKS porous de-icing was provided on leading edges of both fins and tailplanes.

Griffon engines and de Havilland propellers

The Shackleton was powered by four Rolls-Royce Griffon 12-cylinder, vee inline, liquid-cooled, two-speed, single-stage supercharged, fuel-injected engines. Initially, the aircraft featured a pair of Mk 57 engines on the inboard positions and Mk 57A engines on the outboard positions. Each provided 1,960hp for take-off in 'low' gear with +18lb/

sq in boost; or 2,435hp when used in 'high' gear with +25lb/sq in boost using water/methanol injection.

Two three-bladed contra-rotating constant-speed fully feathering de Havilland Hydromatic propellers of 13ft diameter were fitted. The front assembly controlled the rear propeller through a translator unit. De-icing fluid could be fed to the propeller slinger rings from a 33-gallon tank in the starboard wheel bay.

Cross-feeding of fuel between wing fuel tanks was possible, as well as across the fuselage – effectively allowing an engine to run on fuel from any tank. However, each Griffon engine had a separate 26-gallon oil tank. For cold-weather starting an oil-dilution system was fitted, and radiators for both coolant and oil systems were housed in ducts in the forward section of the engine cowlings, which had controllable shutters incorporated.

Each engine was bolted on to pick-up points on the firewall, with a Rotol gearbox mounted behind the firewall in each engine nacelle. Each carried a 6,000-watt, 24-volt DC generator for main services. Engines 3 and 4 featured engine gearboxes driving Dowty hydraulic pumps while engines 2 and 3 featured Hymatic air compressors and Pesco vacuum pumps to drive the flight instruments and bombsight. Two fire extinguisher bottles were installed in each engine nacelle, both of which could be operated manually by the pilots or automatically by inertia crash switches.

Shortly after production of the MR.1 had commenced, the decision was taken to fit the Griffon 57A engine in all four positions, replacing the Griffon 57s that had been in the outboard positions. The designation MR.1A was given to the new aircraft once this modification had been incorporated and all early MR.1s were subsequently upgraded. This got away from the constraints of having to keep two different marks of engine as spares on all Shackleton stations.

Later, the Griffon 58 was introduced on to the Shackleton MR.2 and MR.3 during the Phase 3 upgrades. This variant was effectively a modified 57A with a new auxiliary drive for enhanced power requirements, in which the gearbox teeth were increased from 0.475in (12.7mm) to 0.675in (17.2mm). They were fitted with a two-speed, single-stage supercharger and intercooler, a Rolls-Royce-developed fuel injection pump and a single magneto, mounted above the contra-rotating airscrew's reduction gear housing.

FAR LEFT Looking up into the intake of the No 3 engine on WR963 at Coventry. Clearly visible are the cooling gills of the front radiator. *(Keith Wilson)*

LEFT The de Havilland contra-rotating propeller assemblies on WR963's Nos 1 and 2 engines. *(Keith Wilson)*

RIGHT The galley on Shackleton MR.2/AEW.2, WR963 – which featured an electric grill, pan heater, oven, sink and water heater – was of great assistance to the crews during their long endurance missions. *(Keith Wilson)*

FAR RIGHT The crew rest area located behind the main cabin and ahead of the galley area on Shackleton MR.2/AEW.2, WR963. The couch could be converted by moving the seat back into the raised position, effectively creating two crew rest beds. *(Keith Wilson)*

RIGHT The main cabin was separated from the galley area by this bulkhead door. Just to the left can be seen another two-seat sofa, which also featured a table top for eating at. *(Keith Wilson)*

FAR RIGHT The Elsan chemical closet, as fitted in the rear fuselage of Shackleton MR.3/3, WR977, at Newark. It's not much for a crew of ten (and sometimes more) for up to 18-hour sorties – especially when one considers that one's modesty was spared only by a single curtain around the facility. *(Keith Wilson)*

Accommodation

Accommodation in the MR.1 provided for a ten-man crew consisting of two pilots, two navigators, a flight engineer, and five crewmen with various signalling, gunnery and general tasks, including cooking. The crew comfort aspects took a high priority with the 'standing headroom throughout' feature. Later variants of the aircraft reflected the new roles undertaken and the crew numbers and their positions changed to meet these requirements.

MR.2

The wings and tail of the MR.2 were structurally similar to the MR.1 but featured a redesigned nose and tail. The ASV scanner cupola was removed from the nose and located immediately aft of the bomb bay, and was retractable using two large hydraulic jacks. Three positions could be selected: 'up' when it was practically flush with the fuselage, 'search' when it was extended by around 2ft and 'attack' when the scanner was almost

LEFT Equipment carried on the top of the MR.2/AEW.2 fuselage included (left to right) the observation dome, No 1 VHF (very high frequency) radio aerial, two HF (high frequency) radio aerials and the ARI 18144 Orange Harvest ECM 'spark plug' aerial. *(Keith Wilson)*

LEFT The No 2 heater intake on the port side of WR963. The intake on the starboard side has a larger mouth as it feeds both Nos 3 and 4 heaters. The ram-air intake is the open mouth, and the tube entering from the top is where the exhaust from the heater is routed to provide anti-icing of the intake mouth. The area where the tube enters the intake is double skinned, and the hot exhaust gases are exited behind the curled lip of the intake. *(Keith Wilson)*

LEFT General arrangement drawing of the air bomber and gunner's stations (port side) on board a Shackleton MR.2 (Phase 2). *(Crown Copyright/ Air Publications via Shackleton Preservation Trust)*

F.S.14   **RESTRICTED**   A.P. 101B-1702-1B1, Sect. 1, Chap. 2
                                                                           A.L. 18, June 68

PILOTS' COAMING HEATER CONTROL VALVE
SILICA-GEL CARTRIDGES
GUNNER'S SEAT
HEAD CUSHION
GUN FEED ASSISTER MECHANISM
GUN CONTROL HANDLE
FUSE PANEL
AMMUNITION CONTAINER
FLARE CHUTE CONTROL PANEL
FIRST PILOT'S FOOT WARMER
PYROTECHNIC STOWAGE
TWIN FLARE CHUTE
BOMB SIGHT COMPUTOR
AUTOLYCUS UNIT
FORMER 'E'
STEP
OPERATING HANDLE FOR Nº 1 HEATER RAM AIR CONTROL VALVE
Nº 1 HEATER
AMMUNITION FEED TRACK
GUN APERTURE
AIR TEMPERATURE THERMOMETER
STOWAGE PANEL FOR STOPPAGE TOOL
FIRE EXTINGUISHER
SWITCHBOX FOR Nº 1 BOMB SIGHT
WINDSCREEN WATER TANK
PARACHUTE EXIT DOORS
PNEUMATIC CRATE
STOWAGE BAG
BOMB SIGHT MOUNTING
CONTROL PANEL FOR LOW LEVEL BOMB SIGHT MK.3
AIR BOMBER'S CONTROL PANEL

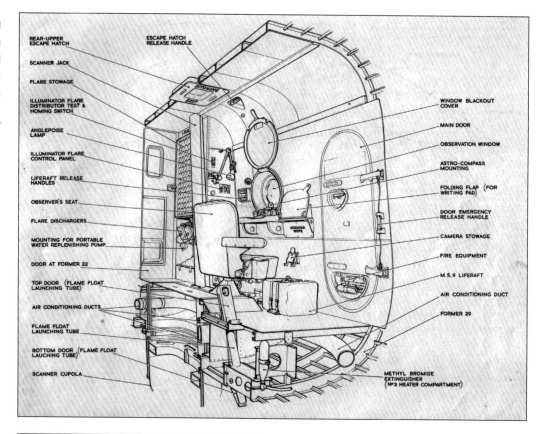

REAR-UPPER ESCAPE HATCH

ESCAPE HATCH RELEASE HANDLE

SCANNER JACK

FLARE STOWAGE

ILLUMINATOR FLARE DISTRIBUTOR TEST & HOMING SWITCH

ANGLEPOISE LAMP

ILLUMINATOR FLARE CONTROL PANEL

LIFERAFT RELEASE HANDLES

OBSERVER'S SEAT

FLARE DISCHARGERS

MOUNTING FOR PORTABLE WATER REPLENISHING PUMP

DOOR AT FORMER 22

TOP DOOR (FLAME FLOAT LAUNCHING TUBE)

AIR CONDITIONING DUCTS

FLAME FLOAT LAUNCHING TUBE

BOTTOM DOOR (FLAME FLOAT LAUCHING TUBE)

SCANNER CUPOLA

WINDOW BLACKOUT COVER

MAIN DOOR

OBSERVATION WINDOW

ASTRO-COMPASS MOUNTING

FOLDING FLAP (FOR WRITING PAD)

DOOR EMERGENCY RELEASE HANDLE

CAMERA STOWAGE

FIRE EQUIPMENT

M.S.9 LIFERAFT

AIR CONDITIONING DUCT

FORMER 29

METHYL BROMIDE EXTINGUISHER (N°3 HEATER COMPARTMENT)

6ft below the aircraft. The selections were made by the ASV operator from his position. A clever safety feature of the scanner was that it automatically retracted when the undercarriage was lowered and could be raised in an emergency using air pressure.

One change that significantly improved ground handling – especially in strong crosswinds – was that the flying controls could be locked for taxiing by a lever that also restricted the throttle movement required. Differential braking was provided through toe-operated pedals on the rudder 'bar', while a parking brake was provided on the port side of the first pilot's instrument panel.

AEW.2

The Shackleton AEW.2 was essentially an MR.2 Phase 3 airframe incorporating increased fuel capacity (up to 3,350 gallons) improved tailplane de-icing and the capability to jettison fuel, along with a strengthened undercarriage. In addition, the ASV-21 scanner and operating equipment were removed and the fuselage cut-out faired over. The bomb bay was shortened by moving the forward bulkhead rearward by around 4ft to provide the space to accommodate the AN/APS 20 radar, which was mounted on to a tubular-frame structure utilising the standard bomb bay pick-up points. The scanner was enclosed in a large cupola similar to that featured on the ASW variant of the Fairey Gannet. However, the new radar installation affected the landing technique of the aircraft. 'Wheeler' landings were no longer possible as a result of the proximity of the base of the radar being very close to the ground. Instead, the aircraft was 'three-pointed' on to the runway, with all three wheels touching the ground simultaneously.

MR.3

Despite its radically different appearance, the MR.3 was substantially similar to the earlier MR.2. The fuselage nose section was altered considerably in layout and incorporated a rearward-retracting, twin 30in-diameter nosewheel assembly and a ventral entrance

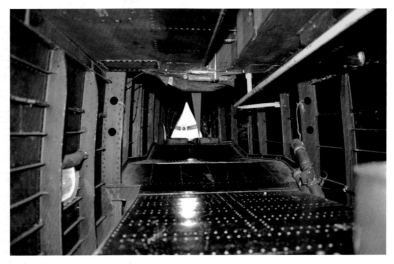

hatch. This change effectively reduced the length of the bomb bay, which was improved by the introduction of pre-loaded bomb beams (see page 110).

Hydraulically operated nosewheel steering was now provided with the controls being positioned on the inside of the first pilot's control column (see page 106). The view from the flight deck was considerably improved with

ABOVE Access to the rear observation position on Shackleton MR.2/AEW.2, WR963, is gained by crawling along the rear fuselage, passing underneath the tailplane. *(Keith Wilson)*

LEFT Shackleton MR.3 aircraft on the production line at Avro's Manchester production facility in 1957 – the height of the MR.3 variant's production. *(Avro Heritage Collection image 696P04591)*

the installation of deeper, wraparound windows and fewer side frames.

The wings were changed in plan form by the introduction of larger ailerons, which were made more effective by spring tabs. Tip-tanks (each containing 256 gallons) were fitted and the internal fuel capacity was increased by using five self-sealing, flexible tanks in each wing. The combined effect was to increase total fuel capacity to 4,248 gallons. Fuel jettisoning was improved, along with modified TKS de-icing equipment.

To move the aircraft's centre of gravity forward of the main wheels, the undercarriage assemblies were radically changed. Electro Hydraulic Limited shock-absorber units were attached to the rear spar and now retracted forward into the engine nacelle. In place of the large single-wheel assembly, each undercarriage leg featured a pair of 49in-diameter wheels with Dunlop Maxaret anti-skid brakes, which were hydraulically operated.

The final and most noticeable change on the MR.3 was to the Phase 3 variant. The fuel capacity was increased to 4,316 gallons and a pair of Bristol Siddeley Viper 203 auxiliary jet engines, each providing 2,500lb of thrust, was added to the rear nacelles of the inboard engines. They provided much-needed additional thrust for take-off and climb, especially when the aircraft was operating at its new maximum weight of 108,000lb. The use of the engines was very limited as the aircraft could not carry any jet fuel required to operate them for any length of time. Instead, the Viper engines used the same Avgas which powered the Griffon piston engines. This factor significantly affected the life of the jet engines.

Interestingly, while the MR.3 Phase 3 aircraft had an improved take-off performance (4,750ft to clear a 50ft obstacle), it had a reduced range and endurance.

MR.3 interior configuration

The Shackleton's two pilots sat in well-upholstered seats set above the standard floor level. As a consequence, their side windows came just about level with their elbows. Vision was more restricted forward, cut off by the long nose. Both pilots had spectacle-type control columns and each had a throttle quadrant and flap levels to their outboard sides. Brakes were on the rudder pedals with trim wheels for all axes mounted behind the throttle quadrants.

Instruments included the standard RAF blind-flying panel plus an ILS (instrument landing system) display. All engine instruments were mounted on a large central panel, and were duplicated at the engineer's position located in the main cabin, just behind the co-pilot.

Between each pilot was a step leading down into the bomb-aimer's area, but they needed to mind their heads on the instrument panel above as they entered the area. There, a couch was located upon which the bomb aimer would lie prone during navigator-directed bomb runs. Above him in the nose was a door which opened on to the nose gunner's seat, one of the best vantage points on the aircraft. The seat swivelled to allow easy access. Once inside, the gunner controlled his twin cannon via a miniature remote control column set in the coaming. There was also a camera mounting position and gunsight.

Behind the two pilots was located the main cabin which ran from just behind the flight deck to a point a few feet aft of the main spar. All equipment was mounted on the port side, except for the engineer's panel. When operating the equipment, each crew member faced sideways sitting in well-upholstered seats designed to reduce fatigue during the long sorties. The seats were able to swivel forward and back for take-off and landing when the crew used the full shoulder harnesses provided.

Immediately opposite the engineer and located just behind the pilot's position was the radio operator, with the aircraft navigator next to him. The next seat was occupied by the tactical navigator (tac-nav), who directed the aircraft during both attack and search operations.

The role of the aircraft navigator was to get the aircraft to a search area and back, while the tac-nav controlled the pattern of a particular search, providing the pilot with courses to steer for both the attack and search phase, presetting the weapons to be used and releasing them at the appropriate time. To assist him in the task he used the projected symbol display

1 Surface movement
   corrector
2 GPI Mk 4C
3 ADF receiver
4 Wind unit Mk 1
5 GPI Mk 1B
6 Remote corrector unit
   No 1
7 Intercom control unit
8 Heading distribution
   changeover switch
9 Oxygen flow indicators
10 Remote deviation
   correction unit No 2
11 Tacan indicator
12 Tacan control unit
13 No 2 compass controller
14 No 2 compass indicator
15 'Blue Silk' indicator
16 Nautical miles counter
17 No 1 compass indicator
18 'Blue Silk' voltmeter
19 GM7 compass controller
20 Bombsight computer
21 Air position indicator
22 Camera doors position
   indicator
23 Camera heater switch
24 Camera doors switch
25 Photoflash dischargers
   master switch

RIGHT The tactical navigator's position on MR.3/3, WR977. His primary role was to track sub-surface contacts throughout the mission. He also had control of the sonobuoys and other tactical weapons and he co-ordinated and controlled the attack.
(Keith Wilson)

1 Plotting control panels
2 Gee Mk 3 indicator
3 Loran indicator
4 Automatic distributor
5 GPI Mk 1C
6 Intercom control unit
7 Indicator Type 7 – Homer RT/SB
8 Sonobuoy repeater indicators
9 Sonobuoy control unit
10 No 3 bombing panel
11 No 2 bombing panel
12 Airspeed indicator (ASI)

RIGHT The first anti-submarine warfare (ASW) position on Shackleton MR.3/3, WR977, is the first or master sonic's position. To the top right is the radio and intercom panel while the main black box (complete with orange display screen) provides the operator – working in partnership with the second sonics operator – with sonobuoy data, allowing them to triangulate the position of an underwater submarine target. This data is passed to the tactical navigator to plot the attack.

To the very left of the image is a small printer that provides the output from a device that can actually detect diesel exhaust from a snorkelling submarine. Essentially an ion spectrometer (IMS), the system was known by the name Autolycus. *(Keith Wilson)*

on a much larger-scale map, and instead of using a geographical chart he had a specially designed graticule chart on the table before him. When being used, the display showed not only the moving aircraft (as seen from above) but the position of the sonobuoys which had been laid, or any other markers that the tac-nav wished to indicate.

RIGHT The signaller's station on Shackleton MR.3/3, WR982/J, at Charlwood. This position is located on the port side of the fuselage, just behind the bulkhead separating the signaller from the first pilot's position. Just to the left is the routine navigator's station. *(Keith Wilson)*

1 Inspection lamp supply plug
2 115-volt AC test equipment supply plug
3 28-volt DC test equipment supply plug
4 Oxygen flow indicator
5 Fuse box
6 Call light
7 Control unit Type 927 (ARI 5848)
8 Intercom control unit
9 Control unit Type 4189
10 Morse key
11 Anglepoise lamp
12 Receiver AD 118
13 HF1 transmitter output switch
14 HF1 supply switch
15 HF2 supply switch
16 Intercom amplifier conference switch
17 Intercom amplifier normal switch
18 IFF master switch
19 IFF I/P facility switch
20 Control unit Type C1128/APX.25 (ARI 5848)

ABOVE The final position in the MR.3/3 cabin is the ASV radar operator's position. This operator has control over the ASV-21 radar's retractable 'dustbin' located under the rear fuselage. This equipment is able to provide tracking of surface vessels and could also be employed during search and rescue missions. To the right is the Orange Harvest radar warning receiver (RWR) equipment, and, to the left, the ASV's operator's station on Shackleton MR.3/3, WR977. (Keith Wilson)

BELOW The lower weapons-aiming position, complete with low-level bombsight, on Shackleton MR.3/3, WR977, at Newark. (Keith Wilson)

1 Crash harness stowage
2 Inspection lamp
3 Scanner position indicator
4 ASV master indicator unit
5 Intercom control unit
6 Call lamp
7 Dimmer switch for panel lighting

8 Dimmer switch for intercom control unit lighting
9 Orange Harvest indicator
10 Orange Harvest control units
11 Bearing marker unit
12 Oxygen flow indicator
13 Air-conditioning louvre
14 Console release handle

ABOVE The 'Blue Silk' Doppler navigation radar mounted on the port side of the rear fuselage, just behind the observer's position on Shackleton MR.3/3, WR977, at Newark. *(Keith Wilson)*

BELOW The ASV-21 radome installation in the 'up' position on Shackleton MR.3/3, WR977, at Newark. *(Keith Wilson)*

BOTTOM By way of comparison, the ASV-21 radome is shown here in the 'down' position on Shackleton MR.3/3, WR982/J, at Charlwood. *(Keith Wilson)*

ABOVE The multiple flare dispensers located on the starboard side, just to the rear of the galley area, of Shackleton MR.3/3, WR977, at Newark. All the chambers are open and the firing pins visible. *(Keith Wilson)*

BELOW The ASV-21 radar and radome were removed from the rear fuselage of the MR.2 aircraft ahead of the AEW.2 conversions and replaced with this radome covering the AN/APS 20 radar located in the front bomb bay, just underneath the cockpit. The large object made landing the aircraft somewhat more difficult and it had to be landed on all three wheels simultaneously, in order to avoid damaging the radar. *(Keith Wilson)*

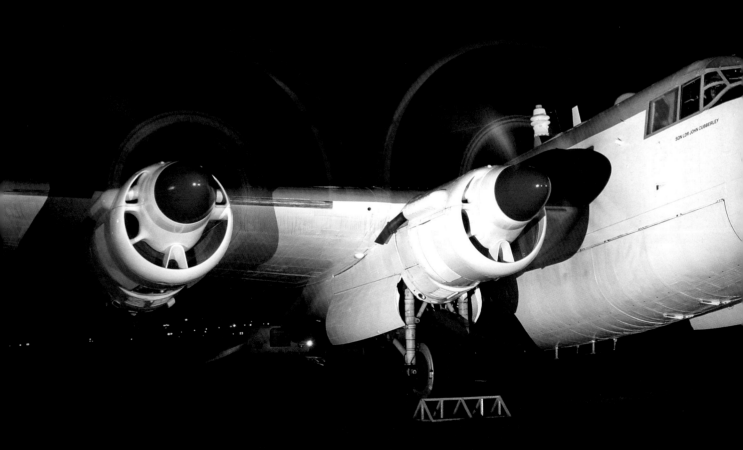

Chapter Six

Operating the Shackleton

━(●)━━━━━━━━━━━━

'Throughout the Shackleton's long and illustrious career, the secret of effective operations was down to the complete harmony and determination of its entire crew. Everyone was totally committed to the task and to each other. It was a memorable experience and a treasure in aircrew lives.'

Squadron Leader John Cubberley, RAF (Retd)

OPPOSITE There may not be an airworthy Shackleton anywhere in the world at present, but at least the Shackleton Preservation Trust frequently run the engines of WR963 and occasionally provide spectacular night-time photographic opportunities, such as this engine run at Coventry in December 2011. The Shackleton Preservation Trust is optimistic about returning WR963 to the air as soon as feasible. *(Tim Badham)*

nacelles to provide additional thrust on take-off, was the current variant in service with Coastal Command. It carried the previous-generation surface and underwater detection equipment; underneath it possessed a capacious bomb bay, while the nose sported a pair of 20mm automatic cannon. Despite its age, the aircraft still packed a punch.

However, the next generation of maritime reconnaissance aircraft – the Nimrod MR.1 – was expected to enter service the following year, initially alongside the Shackleton, but progressively replacing the 'lovely old lady'.

Airborne watch keeper

For the first 19 years of its service career, the Shackleton primarily operated within Coastal Command – in fact it was its workhorse during that period. Its role was that of airborne watch keeper of the seas surrounding Britain and the few remaining bases abroad. Inevitably, that meant a measure of detachment from mainstream RAF life. Indeed, Coastal Command was always more of a 'maritime force' than an 'air force'. On 27 November 1969 that all changed, when Coastal Command was integrated into Strike Command. It meant that there was even closer liaison in the days of increasing unification of command and inter-service collaboration.

By 1969, the Shackleton MR.3 Phase 3, with its two Viper turbojets in the outboard

Flying the Shackleton in Coastal Command

Squadron Leader Mike Rankin (Retired) joined the RAF in 1952, with initial training at the RAF College, Cranwell. Shortly afterwards, flying training commenced and Mike flew the Prentice, Chipmunk, Harvard and Balliol. Now as a pilot officer and with those coveted wings on his chest, Mike was posted to No 8 AFS (Advanced Flying School) at RAF Driffield for a spell on the Gloster Meteor, flying both the T.7 and F.4 versions. Next stop was 231 Operational Conversion Unit (OCU) at RAF Bassingbourn on the Canberra, before being posted to his first tour with 149 Squadron at RAF Gütersloh, on the Canberra B.2. He was later transferred to 59 Squadron, still at Gütersloh on the

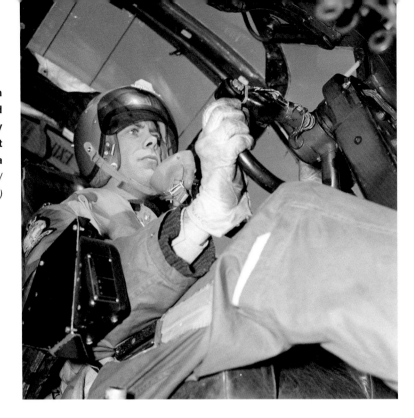

RIGHT A view of the pilot of a 203 Squadron Shackleton MR.2 during a flight covering Iceland and Norway from RAF Ballykelly on 25 May 1964. Both pilots had a well-upholstered seat set above the standard floor level, along with a spectacle-type control column. *(Crown Copyright/ Air Historical Branch image T-4486)*

Canberra B.2 initially, but later on the B(I).8 variant. Mike remained here and completed his tour, having accumulated around 1,000 hours on the Canberra, before his next posting was on the Shackleton where he stayed for the remainder of his service career. From January 1964 to June 1967, Mike was Officer Commanding ASWDU. During his time on the Shackleton fleet, Mike accumulated over 2,500 hours on the 'lovely old lady'.

Mike recalls his time flying the Shackleton:

After a very comprehensive ground school in December 1957, I began pilot training on the Shackleton T.1 and T.4 (identical vehicles as far as handling was concerned) in January 1958. It fitted me perfectly. It was fitted with expensive leather upholstery and the pilots' seats were the most comfortable that I had ever met. I could even stand up in it, despite my 6ft 3in height. Visibility from either of the pilots' seats was exceptionally good and there were three space heaters, though no facilities were ever provided for cooling the aircraft.

BELOW The pilot's centre and emergency panel on Shackleton MR.2/ AEW.2, WR963, at Coventry. Note that on this aircraft, the stores release mechanisms have been deleted and replaced with an engine fire detector test panel. *(Keith Wilson)*

1 Combined heading selector and control unit
2 Bearing and distance indicator – ARI 18107/4
3 Manifold pressure gauge – No 1 engine
4 Manifold pressure gauge – No 2 engine
5 Manifold pressure gauge – No 3 engine
6 Manifold pressure gauge – No 4 engine
7 Rpm indicator – No 1 engine
8 Rpm indicator – No 2 engine
9 Rpm indicator – No 3 engine
10 Rpm indicator – No 4 engine
11 Engine fire detector panel

RIGHT Probably the best-preserved Shackleton anywhere in the world is the former 8 Squadron AEW.2, WR960, currently on display at the Museum of Science and Industry, Manchester. Unfortunately, access to the interior of the aircraft is not permitted, unlike many other UK museums where visitors are invited to join supervised tours, usually conducted by well-informed guides. *(Museum of Science and Industry)*

1 No 1 propeller feathering switch and engine fire warning indicator
2 No 2 propeller feathering switch and engine fire warning indicator
3 No 3 propeller feathering switch and engine fire warning indicator
4 No 4 propeller feathering switch and engine fire warning indicator
5 Dimmer switch
6 Master fuel cock indicator – No 1 engine
7 Master fuel cock indicator – No 2 engine
8 Master fuel cock switch – No 1 engine
9 Master fuel cock switch – No 2 engine
10 Bomb jettison switch
11 Fire extinguisher push switch – port fuel tanks
12 Fire extinguisher push switch – No 1 engine
13 Fire extinguisher push switch – No 2 engine
14 Fire extinguisher push switch – No 3 engine
15 Fire extinguisher push switch – No 4 engine
16 Fire extinguisher push switch – starboard fuel tanks
17 Master fuel cock indicator – No 3 engine

18 Master fuel cock indicator – No 4 engine
19 Master fuel cock switch – No 3 engine
20 Master fuel cock switch – No 4 engine
21 Dimmer switch
22 Call light
23 Cold air louvres
24 ADD stall warning angle of attack indicator
25 Accelerometer
26 Landing lamp switch
27 Landing lamp switch
28 Cold air louvres
29 Airspeed indicator
30 Artificial horizon
31 Zero reader indicator
32 Indicator Type 7 (ILS)
33 Altimeter
34 Gyrocompass
35 Turn-and-slip indicator
36 Radio altimeter
37 Radio altimeter indicator
38 Rate of climb indicator
39 Pilot's centre and emergency panel (see page 103)
40 Engine syncroscope
41 Flap selector switch
42 Flap position indicator
43 Correction card stowage
44 Landing gear selector switch
45 Landing gear position indicator
46 Bomb door selector switch
47 Starboard taxiing lamp switch

48 Bomb door indicator
49 Zero reader indicator
50 Artificial horizon
51 Rate of climb indicator
52 Airspeed indicator
53 Gyrocompass
54 Turn-and-slip indicator
55 Altimeter
56 Warning horn switch, turn-and-slip switch and emergency lighting switch
57 Emergency light switch and turn-and-slip switch
58 Windscreen washer push switch
59 Windscreen wiper handwheel control
60 Port pitot head heater ammeter
61 Starboard pitot head heater ammeter
62 First pilot's aileron control handwheel
63 First pilot's elevator control column
64 First pilot's starboard rudder pedal
65 First pilot's starboard toe brake pedal
66 Second pilot's port rudder pedal
67 Second pilot's port toe brake pedal
68 Second pilot's elevator control column
69 Second pilot's aileron control handwheel
70 P.12 compass

Taxiing

Any multi-engined aircraft can be subjected
to asymmetric thrust, which may be helpful in
manoeuvring on the ground, particularly when a
tight turn is necessary. The effect is very small in
aircraft such as the Comet and the Buccaneer,
where the engines are mounted very close to
the sides of the fuselage, and therefore had too
little leverage to apply torque unless used with
unsafe levels of thrust. Many aircraft with tricycle
undercarriages, including the Shackleton MR.3,
had steerable nosewheels, enabling them to be
driven much like a road vehicle.

Four-engined aircraft like the Shackleton
could exert a strong turning force or torque on
the ground by powering up one of the outboard
engines and even more effectively by using both
of the engines on one side. But the use of so
much power inevitably means that speed tends
to rise, and it may also not be practicable in
confined parking areas because of the powerful
prop wash, so it is not always a suitable
method. But all marks of Shackleton could use
differential braking, applying braking power

separately to the left or right main wheels by the
use of a toe brake on each rudder pedal.

However, the MR.1/T.4 versions had no
internal rudder lock. When taxiing the aircraft in
strong winds, the wind blew the unusually large
rudders of the aircraft away from centre with a
force that could exceed the strength of the pilot.
My own experience of that problem occurred
early in my training when I had not tightened the
lap strap of my safety harness sufficiently. The
rudders blew to the right and despite pressing
hard on the left rudder pedal it moved towards
me, forcing me up off my seat. The right rudder
pedal was then too far away for me to reach its
toe brake to prevent the aircraft from turning.
It accordingly swung left off the taxiway on
to the grass, which – fortunately – was not a
problem at RAF Kinloss. The sandy ground on
the Kinloss airfield was firm and the aircraft did
not sink into it. I was not the first student to
be caught out by the lack of an internal rudder
lock. As the aircraft was undamaged, nothing
was made of the incident and I was able to
store this less-than-satisfactory feature of the

two early marks of Shackleton in my mind without suffering any penalty for it.

The fitting of an internal rudder lock on the MR.2 aircraft completely cured the problems caused by the rudders being subject to the lateral forces of a crosswind. The device itself was simple. With the rudder held in the central position, movement of the rudder lock control inserted a metal pin in the push-rod controlling the movement of the rudders. The device also brought a bar back against the throttles, preventing them from being pushed forward, ensuring that the aircraft could not take off until the rudders were unlocked.

Into the circuit

Once airborne, the aircraft felt just right. If the controls were heavy they did not feel that way because the aircraft was in near-perfect trim. The undercarriage came up smoothly with no noticeable change of trim and when the flaps were retracted I earned a few brownie points by anticipating the small change of trim with the elevator trim wheel. I was now an ex-Canberra pilot, accustomed to a change of trim immediately after take-off that was so powerful as to require not simply a minor movement of a trim wheel, but a major change in the incidence angle of the entire tailplane. I flew around the circuit, momentarily admiring the increasingly wide view of the Highlands south of the field with Findhorn Bay in front of me and the expanse of the Moray Firth to the north.

The after-take-off checks went well and with the aircraft 'cleaned up' I turned downwind, being gently reminded about the speeds we had been taught in Ground School and following the useful points around the circuit that I could use as prompts for the next stage in a visual circuit and landing. Everything seemed to be going my way and I had no difficulty in doing what I needed to do, soon realising that Shackleton speeds gave me a lot more time to get through a circuit and landing than the Canberra had allowed. I was talked through the turn on to finals, reminded again of the speeds and flap settings; particularly the threshold speed – the speed at which I should cross the beginning of the runway – and had no difficulty in achieving the desired speeds at the desired stages of the flight. My first three-pointer was

ABOVE A close-up of the first pilot's control yoke on Shackleton MR.2/ AEW.2, WR963, at Coventry. Note the absence of the additional centre yoke, as steering on the tailwheel versions of the aircraft is provided through the main wheel brakes activated from toe-operated levers on the rudder pedals – along with deft use of the outboard engines. There is no direct mechanical steering of the aircraft as such; the tailwheel is free-castoring. (Keith Wilson)

BELOW The centre overhead panel on Shackleton MR.3/3, WR977, at Newark. Working clockwise from the top left is the row of magneto switches, UHF radios, ADF, ILS, panel light switches, VHF radio and trim panel. (Keith Wilson)

more significantly if the throttles were closed in the final stages of an approach, the propellers would windmill, trying to drive the engines faster. This had the effect – known as 'disking' – of reducing the velocity of the air passing through their diameter and so partially blanking off the rudders and elevators. The elevators in particular would then become much less sensitive and if the round-out before landing had not been completed before the power was cut, it might not be possible to complete it before impact.

I filed that one away, while realising that cutting the throttles completely before rounding-out would be an admission of having lost control and would be very bad airmanship by any standard. Thankfully, I never had to worry about it.

Brakes

The MR.1/T.4 and MR.2 had brakes powered by compressed air. The aircraft's pneumatic system was divided into port and starboard sub-systems driving a series of pneumatically operated services, including special protection for the braking system in the event of a leak. Two storage cylinders at 1,030psi were provided in each sub-system and Nos 2 and 3 engines each drove a compressor to maintain that pressure in their associated sub-system. I do not recall any spate of problems with the brakes.

Cooling air was gathered by vanes in each mainwheel so long as the wheel was rotating. They cannot have had any great cooling effect after landing because of the slow rate of rotation of the wheels while taxiing. If cooling had been required after take-off, the time available for the collection of air would have been virtually nil since the wheels were always brought to a halt by a dab of brakes just after take-off and before they were retracted into the enclosed wheel wells.

Pilots had to be aware that, should brake pressure be lost for any reason, the compressors on each side took a significant length of time to restore it. That situation could arise with any small leak while the aircraft was shut down on the ground, but excessive taxiing or a couple of abandoned take-offs could also drop the pressure below safe levels. This came close to happening at Honolulu, which has two

satisfactory rather than good, but I had showed myself that I could still compensate for the slight inaccuracy in speed control without either setting the aircraft to kangarooing down the runway or thumping down hard. I knew I was going to like the Shack. It was my kind of aircraft.

As time passed I became aware of a few foibles that this big, ungainly looking aircraft possessed – the lack of a rudder lock being just one of them. Another was taught to us, warning one of the consequences of having those contra-rotating propellers. The rudders and elevators were in the slipstream of the two inboard engines. The degree of control exercised by these control surfaces therefore varied with the amount of power used. But

airfields in such close proximity that all arrivals and departures have to take place at Honolulu International. The military Hickam AFB was used by the RAF in transit to Christmas Island and involved roughly a 5-mile taxi to and from Honolulu International. This was something of a challenge to the Shackleton's brake system during the successive phases of Operation Grapple. It must have been an even more serious threat to the tyre temperatures of some heavily laden transport aircraft, but did not seem to be a consideration for the Shackleton.

With the exception of the length of time taken to restore pneumatic pressure in the event of it dropping too low, the brakes of these two marks of Shackleton were very satisfactory in use. They were similar in design to those of much earlier aircraft and so were of a type familiar to many. They were powerful when power was needed and easily regulated, so that cases of tyres being burst by mishandling were rare.

The tricycle undercarriage of the MR.3 aircraft called for a completely different braking system. The brakes were hydraulically powered and operated by toe brake buttons as in earlier Shackleton marks. The well-tried Dunlop Maxaret system was provided, giving very good anti-skid properties. I was a long-term user of Maxaret braking on the Canberra B.2 and B(I).8, and so was familiar with the system. The incompressibility of hydraulic fluid ensured the rapid recovery of temporary pressure losses. It would be difficult to find any complaint about the brakes on the MR.3 aircraft, which made possible the shortest landing runs without undue tyre wear.

In sum, then, the braking systems of all models were satisfactory within reasonable limits, but when taxiing the MR.1 and T.4 variants in strong crosswinds, it sometimes took both pilots to keep the rudders centred.

Flying in Coastal Command

It was with much relief that I finished that tour with RAF Germany and left for Coastal Command and training on the Shackleton. What I found delighted me. Flying had a purpose again; morale was high because development of our professional skills was encouraged by giving us incentives to improve. Competition was everywhere – inter-crew, inter-

LEFT The large rudder pedals at the co-pilot's position on Shackleton MR.3/3, WR977, at Newark. Also visible in this view are the toe-operated brakes just beyond the foot straps. (Keith Wilson)

LEFT By way of comparison are the first pilot's rudder pedal (red) and toe-operated pedals (black) on Shackleton MR.2/AEW.2, WR963, at Coventry. (Keith Wilson)

squadron, inter-group – with minor prizes such as a mention in the quarterly magazine or the printing of the best reconnaissance photograph along with major prizes such as silver cups presented by donors for various aspects of maritime operations.

Digital simulators had not yet come along, but ingenious crew procedure trainers were beginning to supplement less detailed methods of training crew and having a considerable effect on the competence of crew from the moment of completing the Shackleton conversion course. For those who needed the added incentive, Coastal Command's Categorisation Board examined the operational efficiency of everyone, in great depth, every year. Traditionally, in peacetime, life in the RAF

ABOVE Shackleton MR.2 Phase 2 aircraft (WG533/B, WL757/C and WL755/T) of 224 Squadron at RAF Gibraltar, being prepared for a sortie at the airfield in March 1962. No 224 Squadron was participating in NATO Exercise Dawn Breeze at the time of the image and the groundcrew can be seen moving depth charges towards the bomb bay. *(Crown Copyright/Air Historical Branch image PRB-22433)*

MR.2's requirements at the point of its fully stalled landing technique – the three-pointer – now amazes me and, perhaps, goes some way to explaining the trepidation I felt when at RAF Ballykelly one night I turned down a request from Air Traffic Control to land the nosewheel-fitted MR.3 aircraft slightly downwind.

RAF Ballykelly had a problem that had been troubling them all day. There was an as-yet undiagnosed fault in the power supply to the runway lights. Whenever they switched the lighting on the main east–west runway from one direction to the other, the power failed and the airfield was flung into darkness. 'Would I be prepared to land downwind so that they did not have to reverse the lighting again?' was the question posed to me.

The tailwind was very small and I would have accepted it without a qualm had I been flying an MR.2. Quite why I was worried enough even to ask them how long it had taken to restore the power now surprises me. And when they said 'About 15 minutes', I replied that I would prefer to land in the 'correct' direction and had enough fuel to hold for much longer than 15 minutes.

It is easy to say now that I did not need to do that, but it didn't feel wrong at the time (which of course was when it counted). The runway lights stayed on after the reversal, but failed just as I turned off the runway. That was the result of an of airmanship decision that, even though I still feel a bit mean because I caused a certain

for aircrew could be described as 'cushy', but these high levels of professional activity in Coastal Command squadrons were much more satisfying and our training was very thorough.

And now is the time to consider what I found to be the major difference when flying the heavier and more modern-looking Shackleton MR.3. By that time I had roughly 2,000 flying hours in the tail-dragging MR.2, but only around 50 hours on the MR.3. My sensitivity to the

RIGHT Armourers from 224 Squadron loading a rack of six depth charges into the bomb bay of Shackleton MR.2, WG533/B, at RAF Gibraltar. *(Crown Copyright/Air Historical Branch image PRB-22437)*

FAR RIGHT Once the depth charges have been loaded, a pair of light series carriers is added to the bomb bay, positioned on either side of the depth charges. *(Crown Copyright/ Air Historical Branch image PRB-22436)*

amount of inconvenience to others, I know to have been the right decision. It ensured that there was nothing different about the landing and in which I therefore had full confidence. It is a minor matter among many more important such matters that helped me fly almost 10,000 hours without bending an aircraft.

Operating the systems

Squadron Leader (Retired) John Cubberley joined the Royal Air Force in 1953. He completed his initial, basic and advanced radio training at RAF Swanton Morley before completing Gunnery School at RAF Leconfield. Next stop was the Maritime Reconnaissance Squadron (MRS) at RAF St Mawgan before joining the Shackleton Operational Conversion Unit (OCU) at Kinloss. His first posting was – as a sergeant air signaller – to 269 Squadron at RAF Ballykelly, but he spent the first six weeks on the beach doing dinghy training!

John recalls:

At the time, everything was rationed – including aviation fuel. At the end of the month a new allocation of Avgas arrived and we were back in the flying game. The trips were long and hard and the MR.1 or MR.1A wasn't particularly comfortable – it was like the 'black hole of Calcutta'! That said, it did the job extremely well. In 1954, we were up against the diesel-

ABOVE A clearly posed image of the crew performing a pre-flight check on an unidentified 120 Squadron Shackleton MR.3 (coded 'A') ahead of the NATO Exercise Fairwind Seven at RAF Kinloss in June 1962. The crew entry steps are visible in this image, just behind the nosewheel undercarriage doors. *(Crown Copyright/Air Historical Branch image PRB-23210)*

BELOW LEFT The first pilot's throttle quadrant on Shackleton MR.3, WR982/J, at Charlwood. The top four levers operate the throttles on each engine while the fifth lever provides a friction lock to prevent them inadvertently moving. Below are the four propeller pitch control levers, with a friction lock to the right. To the top is the redesigned rudder lock mechanism. *(Keith Wilson)*

BELOW The first pilot's throttle quadrant on the Shackleton MR.2/AEW.2, WR963, at Coventry is very similar. The additional red lever at the top is the rudder lock. *(Keith Wilson)*

electric-powered submarines of the Soviet Navy – nuclear-powered subs were a long way off. We were using the ASV-7 radar in the Lancasters at the time, the Shackletons had the ASV-13 and in the 1960s the ASV-21. Our radar tactics were considered and effective and we maintained a tight control of Soviet movement through our areas of responsibility.

During my stay at Ballykelly, we did some of the trooping flights into Cyprus during the Suez crisis in 1956 and was also involved in the great adventure to Christmas Island in October 1957. One of the interesting aspects of that deployment was the journey from Sacramento, California, to Hickam AFB on Hawaii. After 10 hours being airborne, we passed the 'point of no return' and with no alternative airfield we simply *had* to find our destination! Thankfully, land appeared on the radar at around the anticipated time and we landed safely at Hickam. It was a new experience but soon forgotten with the contemplation of four days off in paradise.

Once on Christmas Island, our first objective was to keep people out of the test area and we flew daily missions to ensure this plus a series of special meteorological profiles running up to each test day. The test involved the detonation of four bombs, two on the ground mounted on towers, and two free-fall from the air. Our crew was airborne for both of the first two tests and we saw the explosions from a distance. For the first live free-fall Hydrogen bomb test, our crew was scheduled to be on the ground. I remember the heat of the sun burning the side of my face but the heat of the explosion was hotter on my back! I remember the ball from the explosion forming shortly after the explosion at 10 o'clock in the morning and the ball was still growing at 4 o'clock in the afternoon. Our job that day was to take off 1 hour after the explosion and fly over 'ground zero' at 250ft. After getting airborne in the MR.1, we climbed up to 6,000ft, went into a dive and got the speed up to around 310; and with everything shaking we flew across 'ground zero' absolutely flat-out! You could have put a sixpence on to the top of the pile. The ground had been pulled up by the funnel effect of the explosion into a pile around 100ft high. If the explosion had been detonated at ground zero, rather than

7,000ft up, it would have sucked the whole island up into it. I was airborne for the final test and once again, saw it from a distance.

My tour with 269 Squadron lasted almost 5 years before I moved on to MR.3s with 201 Squadron at St Mawgan.

Later, I was selected for the AE (Air Electronics) course and completed it by 1960. Shortly afterwards I was commissioned as a pilot officer and posted to 120 Squadron operating the MR.3 Phase 1 aircraft. After studying for a while I was promoted again, this time to flying officer. I completed the tour before being moved to Malta with 38 Squadron, having just been promoted again to flight lieutenant. We were a good squadron and we all had a great life on Malta. We worked very hard, with lots of deployments including Akrotiri (during the EOKA problems), flying up to 120 hours a week on some occasions.

Around the time, g-meters had been fitted to the aircraft, positioned around the rear spar on each aircraft. Normally, we would 'pull' around 10 points on the g-meter during a typical maritime sortie but on one occasion we managed to pull 1,000 points on the gauge. That night, the wind was from the north, over the Aegean Sea, and it made the air mass tumble. It was so bumpy! Everyone on board suffered with air sickness apart from the first pilot and me. I spent most of that trip standing over his shoulder keeping him distracted from the turbulent weather battering us around. That was a rough trip!

I had managed to get myself an 'A' Category professional rating which resulted in my selection to the Coastal Command Categorisation Board, based at HQCC (Headquarters Coastal Command) Northwood, and travelled to squadrons throughout the Command, examining individuals and crews. This was followed by a 2½-year desk tour at HQCC administrating posting and careers of AEO officers and all airmen aircrew within the Command.

Then it was back to flying with another 2½-year exchange tour with the Royal Canadian Air Force (RCAF) at Greenwood, Nova Scotia, operating on their Argus 1 and 2 aircraft. Acceptance of the posting unfortunately restricted the pattern of promotion I had fortunately achieved, but the benefit of such a treasured opportunity to serve with the Canadians, and to take my family with me, was well worth it.

I returned to RAF St Mawgan where I took up a post of OC Maritime Acoustic Analysis Unit (MAAU). The MAAU operated in direct support of the Nimrod. It provided briefing support and after-flight analysis for the new frequency acoustic system now installed into the Nimrod MR.1 aircraft. This was followed by a 1-year period as a ground instructor with MOTU, followed with the Nimrod OCU and a 1-year period with 120 Squadron at RAF Kinloss.

This short-term posting was followed by another exchange posting, to the US Navy Facility (NavFac) at Brawdy, where I worked on low-frequency acoustics in the ocean. NavFac Brawdy supported mobile operations via Maritime Operations and my role became that of regional watch officer.

After I completed that tour, I was transferred to the Admiralty Research Laboratory, Teddington, with the Joint Acoustic Analysis Centre (JAAC). Here, I served as the analysis officer for the composite maritime fleet of Nimrod MR.1 and MR.2 aircraft. It was a tremendous role, not just analysing and reporting on the activities of the Nimrod aircraft but also on the submarine activities too. Towards the end of this tour I was promoted to squadron leader and briefly assumed the role of OC JAAC before being 'invited' to take up a post of Defence Analyst under the Director of Scientific and Technical Intelligence, a group made famous by R.V. Jones in the Second World War. My post was in DI57 Navy and I served in this role for 5 years before retiring from the RAF on my 55th birthday.

When I left the RAF I had around 6,500 hours in my log book; more than 4,000 hours on the various marks of Shackletons, along with 2,500 hours on Nimrod and Argus aircraft.

In the Shackleton, my job was getting the best from the tight team of enthusiastic AEops and air sigs who had so many tasks to complete during operations; and which contributed significantly in bringing successful prosecutions. The following are my recollections of the activities of the crews on the MR.1, MR.2 and MR.3 aircraft during the almost 13 years I spent on the 'mighty beast'.

BELOW A great view of the operating stations just behind the flight deck of Shackleton AEW.2, WR960, on display at the Manchester Museum of Science and Industry. To the right is the engineer's position, while to the left is the signaller's station. (Museum of Science and Industry)

RIGHT The engineer's panel on Shackleton MR.3, WR982/J, at Charlwood. The major difference between the engineer's panel on the MR.3/3, compared with that on the MR.2/AEW.2, is the additional controls required for the two auxiliary Viper jet engines. On this MR.3/3 aircraft, the controls are cleverly located under the Perspex shelf. Access is gained by simply lifting the shelf upwards. *(Keith Wilson)*

FAR RIGHT On the Shackleton MR.2/AEW.2, WR963, at Coventry, the engineer's shelf is solid, without any controls located beneath it. *(Keith Wilson)*

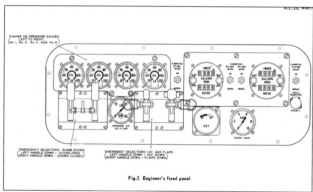

Fig.7. Engineer's fixed panel

LEFT A diagram showing the engineer's fixed panel as fitted to a Shackleton MR.2 (Phase 3). *(Crown Copyright/Air Publications via Shackleton Preservation Trust)*

BELOW LEFT A diagram showing the engineer's main panel as fitted to a Shackleton MR.2 (Phase 3). *(Crown Copyright/Air Publications via Shackleton Preservation Trust)*

BELOW A diagram showing the engineer's auxiliary panel as fitted to a Shackleton MR.2 (Phase 3). *(Crown Copyright/Air Publications via Shackleton Preservation Trust)*

Fig. 8. Engineers main panel

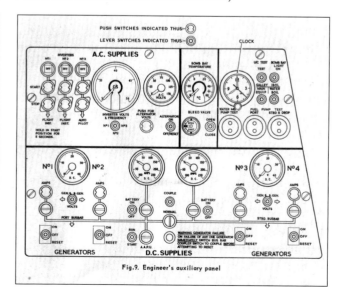

Fig.9. Engineer's auxiliary panel

The radio operator

Regardless of where you had to go, the radio operator had an important job to do – keeping the aircraft in contact with operational control. The initial equipment on board the early MR.1 Shackletons was the former wartime Marconi 1154/55 radio equipment. It wasn't until 1958 that the SDR18 was installed, and later still that we got the Collins equipment. The old equipment had a maximum transmitting output of just 80 watts; it was nothing like the modern SSB (single side-band) equipment.

The radio operator has to be concentrating the whole time. The sending and receiving of coded messages – long strings of numbers and letters, using on average groups of 30–40 – often took up to 30 minutes to get through or receive. Under target-attack situations, the operator went into overdrive, during which he had to maintain total control. Concentrating at that level on sorties of 8 hours or more was a most demanding task.

The AEops air signaller

The AEops air signaller operated the radar, 20mm cannon, acoustic sets, cameras, ordnance and also manned the lookout positions. To be effective, you could not spend more than 45 minutes on the radar screens without a reasonable break. This was achieved by setting up a roster which considered this problem and at least rested the eyes from the rotating radar screen for a period. It was not an opportunity to rest, though, because there were many more tasks for the crew to cover. Consequently, all the crew was always busy.

To be a good lookout required training. The ocean is a dynamic place and there is a lot of it! You had to be trained on just how to 'look out' to be able to see very small objects. At night the human eye relies almost totally on photo receptor cells (rods) and methods to maximise this were adopted as much as possible. Long periods on the lookout stations were inevitable and could be extremely tiring and after a while, your eyes deteriorate and become ineffective. The two pilots assisted greatly with the lookout task, allowing them to instantly commit the aircraft to a visual attack. Time was always of the essence.

LEFT The navigator at work aboard a 206 Squadron Shackleton MR.3 during NATO Exercise Dawn Breeze out of RAF Gibraltar in March 1962.
(Crown Copyright/Air Historical Branch image PRB-22427)

LEFT One of the radar operators aboard a Shackleton T.4 operating the ASV-21 radar system during Exercise Unison in 1963.
(Crown Copyright/Air Historical Branch image T-4153)

LEFT A signaller on board a 206 Squadron Shackleton MR.3 occupies the port beam blister window. Standard equipment was a pair of good binoculars and the trusted 'Mk 1 human eyeball'.
(Crown Copyright/Air Historical Branch image PRB-22428)

LEFT The signaller's role also included that of photographer, as seen here. Note the port beam blister window in the open position, with the screen lifted up and out of the way.
(Crown Copyright/Air Historical Branch image PRB-22429)

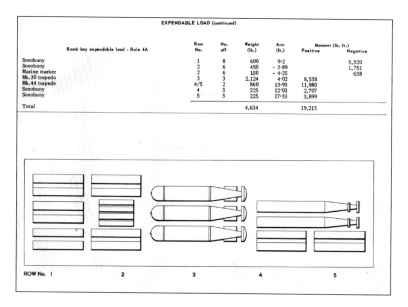

| EXPENDABLE LOAD (continued) | | | | | | |
Bomb bay expendable load - Role 4A	Row No.	No. off	Weight (lb.)	Arm (ft.)	Moment (lb. ft.) Positive	Moment (lb. ft.) Negative
Sonobuoy	1	8	600	9·2		5,520
Sonobuoy	2	6	450	- 3·89		1,751
Marine marker	2	6	150	- 4·25		638
Mk.30 torpedo	3	3	2,124	4·02	8,538	
Mk.44 torpedo	4/5	2	860	13·93	11,980	
Sonobuoy	4	3	225	12·03	2,707	
Sonobuoy	5	3	225	17·33	3,899	
Total			4,634		19,215	

ROW No. 1   2   3   4   5

ABOVE A diagram showing a 'typical' Maritime Reconnaissance expendable bomb bay load, as used on a Shackleton MR.2 (Phase 3). The weapons are carefully located and released in order to minimise any CofG (centre of gravity) issues with the aircraft's handling characteristics. The various weapons and their weights can be seen in the table above. *(Crown Copyright/Air Publications via Shackleton Preservation Trust)*

BELOW A diagram providing an interesting comparison with the typical Maritime Reconnaissance load above. Here, the expendable weapons load on board a Shackleton MR.2 (Phase 3) includes one single AS bomb in row 3. While the capability was available, this weapon was an atomic-store anti-submarine device, nicknamed 'Shape', that was never actually used by a Shackleton operationally. *(Crown Copyright/Air Publications via Shackleton Preservation Trust)*

| EXPENDABLE LOAD (continued) | | | | | | |
Bomb bay expendable load - Role 1A	Row No.	No. off	Weight (lb.)	Arm (ft.)	Moment (lb. ft.) Positive	Moment (lb. ft.) Negative
Sonobuoy	1	8	600	- 9·2		5,520
Sonobuoy	2	6	450	- 3·89		1,751
Marine marker	2	6	150	- 4·25		638
Mk.30 torpedo	3	2	1,416	4 02	5,692	
A.S. bomb	3	1	1,185	4·02	4,764	
Mk.44 torpedo	4/5	2	860	13·93	11,980	
Sonobuoy	4	3	225	12·03	2,707	
Sonobuoy	5	3	225	17·33	3,899	
Total			5,111		21,133	

ROW No. 1   2   3   4   5

Underwater targets

When on patrol and you have to make an attack, you have to broadcast on the radio what you are doing – a 'what, where and when' coded message – which will be transmitted on HF (high frequency) as well as on the standard U/VHF frequencies, so that friendly ships in the area are made aware and can act accordingly.

The radar operator will have made contact with the target and will have to use all of his skills to keep hold of that contact. If you want to approach the contact without him being aware of your position, you need to change into 'standby' mode and mark the contact on the screen. Sadly, the original ASV-7 equipment wasn't very good at this. The ASV-13 wasn't a great deal better but the ASV-21 was quite brilliant at it – by providing a tactical manipulation of the data. During this type of attack, 'clutter' was the big problem on the radar screen so it was required to position the aircraft so that radar had a downwind aspect, thereby reducing the difficulty.

Throughout an attack, the submarine was always at an advantage. When the Shackleton is transmitting a radar signal that has to go all the way to the target and then all the way back again – effectively travelling twice the distance – it can let the submarine know that you are looking for him and, more importantly, where you are. A disappearing radar contact attracted the aircraft to fly to the last known position of the possible submarine and deploy a Mk 1C passive sonobuoy, which had a 2,000yd range.

The passive sonobuoys were listening for cavitational noise (small liquid-cavitation-free zones ('bubbles' or 'voids') that are the consequence of cavitational forces acting upon the cavitational liquid) which would become apparent if the submarine attempts to escape using high power settings. The active sonobuoy, meanwhile, was emitting signals that would identify the location of the target. Once the target location is clearly identified, a torpedo could be deployed.

In the early days of Shackleton operations, these were usually the Mk 30 passive torpedoes, which were utterly useless and soon withdrawn from service! They were replaced by the US-manufactured Mk 44. When the Mk 44 was first tested against a submarine, the ship couldn't get away from it and it went straight through the conning tower and lodged there!

Most of the time we were exercising against Royal Navy and some friendly submarines but our main role was to track Soviet submarines. We couldn't drop weapons against them – that would have caused a war – but once we made 'contact' with them, they would usually dive and try to get away from us. We knew how far they could travel before having to come back up to the surface to recharge their batteries; at which point a large number of Shackletons would be in the search area looking for them. If we found one it would have to dive again and that was really the name of the game – keeping them down, well below the surface. It is much the same strategy used against the U-boats during the Second World War; you didn't necessarily need to sink them, just keep them submerged, thus limiting their operational deployment. In war, that meant preventing them from engaging with the Allied convoys, because they could not use the necessary speeds to achieve an engagement position.

While at RAF Ballykelly, St Mawgan and Kinloss, we operated against the 'Foxtrot', 'Romeo', 'Whisky' and 'Zulu'-class diesel-electric submarines that were usually transiting 'our' waters, en route to the Mediterranean. Once there, the RAF Gibraltar and RAF Luqa, Malta-based squadrons would monitor them. Occasionally, the target was a 'Golf'-class submarine, equipped with missiles within its conning tower. These were usually en route to the west Atlantic to deploy into a location where their guided missiles could operate. In the early days, these had a range of only 500–600 miles, so they had to be taken close to theatre, normally the coastline of the United States. When the Soviets developed long-range missiles, they could have been launched from pretty much anywhere in the west Atlantic. During the Cuban missile crisis of October 1962, the Shackletons were carefully monitoring the submarines en route to theatre. It was a time of very high alert; we really thought there was going to be another war!

On-the-water targets

As part of the Maritime Reconnaissance role, the Shackletons would go out and find Soviet Navy vessels, before identifying and monitoring their activities. Occasionally, these included the Soviet intelligence-gathering vessels (AGIs – intelligence trawlers) that would occasionally encroach inside British territorial waters. They were especially keen to sit off the British coastline, adjacent to armament ranges where they could monitor missile testing. One of their favourite haunts was off the coast of the Hebrides and they provided regular targets for Shackletons.

'Crossing the coast' checks

Shortly after leaving the airfield, the aircraft would cross the coastline en route to the planned area of operation. 'Crossing the coast' checks would have been completed by the whole crew. This extensive checklist included such unlikely items as cameras and binoculars being in operational readiness, as well as checking the necessary electronic gear. Every Shackleton carried all the necessary equipment to conduct a visual airborne search operation, and at any time was liable to be called to join a search for a ship or an aircraft in distress. All maritime stations maintained a duty search and rescue Shackleton and crew on standby, together with a second aircraft and crew on a lower state of readiness on a squadron rotational basis. It was also the time that all flares were loaded in the tubes including photoflash flares from the roof-mounted discharger for the attack photographic record.

Sonobuoys and flares

The sonobuoys and recce flares were dropped from the aircraft bomb bay. Location flares

BELOW Armourers preparing to load an active sonobuoy on to a 120 Squadron Shackleton MR.2 at RAF Kinloss ahead of NATO Exercise Fairwind Seven in June 1962. To the right of the trolley (numbered 13, 9 and 4) are passive sonobuoys, while to the left of the trolley are active sonobuoys.
(Crown Copyright/Air Historical Branch image PRB-23207)

were fired from tubes mounted in the side of the aircraft aft of the wing and were projected upwards. In order that a rapid sequence of flares could be fired, the tubes were in multiple banks. They were used during night operations when the close barrage of flares could be fired to illuminate a patch of water up to a mile in length with brilliant clarity, to enable the pilots to carry out a visual attack.

The fighting part of the aircraft

Throughout the Shackleton's long and illustrious career, the secret of effective operations was down to the complete harmony and determination of its entire crew. Everyone was totally committed to the task and to each other. It was a memorable experience and a treasure in aircrew lives.

Nuclear-powered submarines spell the end for the Shack

In the mid-1960s, nuclear-powered submarines started to appear. The Shackleton was not equipped to meet this challenge, and the need to bring into service the advanced technology required to match this threat became of paramount importance. Sadly, the 'old lady' had to go; and in 1968 the mighty Nimrod MR.1 entered service.

By a quirk of fate the Shackleton days were not completely over as government changes pressed the RAF to look for a replacement aircraft for the Fleet Air Arm (FAA) Gannet. The result was that 12 Shackleton MR.2 aircraft were selected for a refurbishment programme

that would equip them as Shackleton AEW.2 aircraft. They were to be operated by 8 Squadron, a former jet fighter squadron that had operated alongside Shackleton squadrons in Aden. The Shackleton career did not end until 1991, when it completed 40 years of front-line service – a brilliant and proud record.

Starting and running WR963

In 1998 Rich Woods unwittingly climbed aboard an aircraft that was going to play a big part in his life. He was 16 years old at the time and the aircraft he climbed aboard was an old, faded and seemingly dead Avro Shackleton languishing on the edge of Coventry Airport. At the time, he was keeping himself amused by exploring it while waiting for a flight on a DC-3 – which had gone unserviceable. Hours later he had to be prised out of the Shackleton in order to catch his ride in the DC-3. His last memory of that day was wishing he could do something to help the old aircraft; little did he realise he would get his wish.

Fast-forward to 2009 and a chance remark to a friend resulted in him arriving back at Coventry Airport armed with spanners and enthusiasm, and backed up by his father (Dave Woods) to join in the work of the Shackleton Preservation Trust . . . a Trust dedicated to restoring that very same Shackleton. The biggest thing he had restored thus far was a classic Jaguar, and his career had taken a distinct turn away from aviation in that he was working for the Civil Service. However, he was willing to learn and the manuals were easy to photocopy.

With dwindling numbers of Shackleton flight engineers available, he was fortunate enough to be chosen by the SPT to take on the role of engineer for ground runs. His first time starting WR963 was in May 2010 and it still remains a fantastic experience to bring the four Griffon engines to life. His role within the Trust expanded and he has now headed up the maintenance on the aircraft for the past 5 years; a job which has gained an interesting new angle with the renewed work to return the aircraft to flight.

Who better to describe the pre-flight inspection, engine start-up and running of WR963? Rich takes up the story:

Pre-engine run inspection

WR963 stands idle for at least four weeks between ground runs; as a result it must be checked thoroughly before the engines are run up. We start with the simple things such as making sure that there is no water collecting in the fuel tanks, by drawing a sample from each of the drains on the Nos 1 and 2 tanks. The Nos 3 and 4 tank groups (they are interconnected) on WR963 are not used due to the age of the flexible rubber fuel cells.

A water check is also done at each priming pump – of which there are two, one for each wing – located in the forward end of each landing gear bay. This is the lowest part of the fuel system and water collecting here can corrode the pump internals. A seized pump will blow the fuse in the aircraft rendering any further attempts at priming futile.

While in the starboard undercarriage bay, the hydraulic hand pump handle is checked to make sure it is secured and its pin is in place preventing movement.

On walking around the aircraft, the various covers and blanks are removed from pitot heads, heaters and intakes, and the chocks are checked that they are snug against the tyre and not likely to slip free unintentionally. We make sure that the undercarriage ground locks are firmly in place; four large, sprung items for the main landing gear, and a collar with safety pin at the rear to lock the tailwheels down. While checking the tail, the towbar can be removed if required, though unless we are taxiing the aircraft we tend to leave it in as it's a heavy item to move around on the grassy area we are parked on.

Moving on board the aircraft we remove the control locks from elevator and aileron and stow them in their clips in the beam of the aircraft. After a brief check of the switches on the engineer's and pilot's panel to make sure nothing is out of place (we don't like surprises) we connect the aircraft's internal batteries and check voltage, fuel levels across the tanks, entering the data on to the log sheet.

In the pilot's position, we check the gauges near his left knee that indicate pneumatic pressure; most important as this indicates how much we have available prior to engine start for brakes and engine services. We do suffer the

odd leak, so have a small reserve of pneumatic pressure available from the redundant storage crate in the extreme nose of the aircraft.

The outside air temperature (OAT) is noted and logged as this dictates how long each engine must be oil primed before starting, with 4 minutes at 0°C being the most, removing 1 minute off the required time for every 5° above zero. Oil priming does not have to take place if the engines have been started within the last seven days.

If everything is satisfactory, the batteries can be switched off, and we can proceed.

Ground equipment

Attached to the front of the aircraft by a large cable is a 24-volt DC power supply, commonly known as a Trolley Accumulator or 'Trolley AC'. Originally full of large lead acid batteries, it now houses a deceptively small modern battery pack which provides enough punch to start a cold Griffon at sub-zero temperatures.

Fire extinguishers are always to hand, and in the absence of a large bottle and lance we make use of Coventry Airport's fire cover for most of our engine starts. Occasionally, the Shackleton may suffer a priming fire due to the coarse method it uses, with WR963 having a couple of memorable ones over the last few years. Thankfully, for all the spectacular flames they are relatively harmless and by keeping the propeller turning and shutting off the fuel they are usually extinguished as rapidly as they appear.

In a large trolley parked near the starboard main undercarriage can be found four large cylinders

LEFT Before any engine run is conducted the fuel is checked for the presence of water. A small amount is drained from tanks 1 and 2 on MR.2/AEW.2, WR963, at Coventry and examined for the presence of water, which appears as small droplets at the very bottom of the fuel. If water is present in the fuel tanks, it will fall to the bottom and collect around the drain plug. If any sign of water is found during this exercise, the tanks are drained until no further water is found. The No 3 and 4 tank groups (they are interconnected) on WR963 are not used due to the age of the flexible rubber cells in these fuel tanks. *(Keith Wilson)*

of nitrogen – the Shackleton has quite a thirst for this gas, with not only the usual uses for tyre and oleo inflation, but also to charge the pneumatic system's four storage cylinders up to 1,030psi.

Prop 'dance'

Normally seen on radial engines, we have adapted a process to pre-oil each engine on the Shackleton. The Griffon has an inherent weakness in that it suffers from excessive camshaft wear if the engine is run infrequently for short periods – much as we do with our ground-running of the aircraft.

Using the redundant oil dilution system on the Shackleton, onboard pre-oiling pumps were fitted and plumbed in permanently to the engine's oil priming connection. By switching on the pump, fresh oil can be pumped up to the camshafts and around several other critical spots in the engine before it is turned over for starting. Rotating the engine by hand (engine stone cold, no priming or fuel, and ignition switches 'off' and caged) while the pump is running, results in sufficient oil at all points of the engine within 4 minutes at 0°C OAT (outside air temperature).

Crew and briefing

Crew briefing is done by the aircraft captain prior to start, which is normally the pilot. He will detail what the day's exercise is, the duration and whether there are any special circumstances (such as taxiing, or if there has been a major component changes and things to consider). Individual duties are assigned such as a lookout on each wingtip, the ground engineer, and a headcount is taken as to how

many people will be on board the aircraft during the run-up.

Minimum crew on board are pilot, co-pilot, engineer and beam, with at least one team member covering the duties of ground engineer. The airport's Air Traffic Control is contacted at this point to advise them of the time of our engine start, as we don't want to alarm them with unscheduled movements.

The start routine

Once all the pre-start checks have been completed by each crew position – things like checking for full and free movement of controls, that all hatches and windows are secure – the 'power on' checks are carried out. The No 1 and 2 fuel tank cocks are opened along with the master cocks for all four engines. The indicators on the engineer's panel move from crossline to inline, providing a visual indication as to where the fuel can flow.

The aircraft's internal batteries are switched 'on' via the overhead electrical panel, and the voltages checked. The large yellow 'Generator Failure' light will now be flashing as the aircraft's systems are indicating that all four are offline – which is correct as the engines aren't running.

The first engine selected for starting is No 3, so the start master switch is set to 'on' and the large rotary switch for engine priming is set to No 3 engine. As the priming pump runs, a small yellow light illuminates to indicate the correct fuel pressure is reached.

When clearance to start is received from the ground engineer, the pilot turns on the ignition switches, and gives the order 'Turn No 3!' At this point the engineer presses in the priming switch to feed priming fuel to the selected engine, and at the same time holds the switches for the boost coil and starter to 'on'.

The engine starts to turn and now you are looking at the propeller for indications that the engine has caught, the exhaust pipe to monitor for over-priming (black smoke, or worse a flame), with the odd glance at the main instrument panel to watch for a tell-tale flicker on the engine rpm gauge.

At this point there is often the temptation by those in the pilots' seats to try to catch the engine as it starts to cough. This can be an irritation to the engineer having to get them to

ABOVE Selecting the fuel tanks ready for engine starting. In this case, the port No 2 tank cock is being switched to 'open'. The 'doll's eye' indicator will go from crossline (closed) to hatched, while the cock is moving, to inline (open). Fuel tanks selected for the ground runs on WR963 are normally port and starboard Nos 1 and 2 tanks, with cross-feeds all closed. *(Keith Wilson)*

ABOVE Rich Woods occupying the engineer's seat in WR963 ahead of an engine run at Coventry on 10 September 2014. Here he is priming the fuel lines prior to engine starting, in order to purge any air out of the fuel lines that has accumulated since the engine was last shut down. This action must be carried out on all four engines. The buttons being held – in order – are: No 1 (port or starboard depending on which engine is having its lines primed), fuel tank cock 'open', fuel master cocks 'on' for the selected engines on the pilot's panel, slow running cut-off switches held 'on', and No 1 boost pump 'on'. The boost pump is run for 45 seconds to prime the line before being switched off. The switches are all returned to 'off' or 'closed' in reverse order. *(Keith Wilson)*

BELOW Starting the engine (in this case No 3). Actions prior to this will have been carried out on the pilot's panel and overhead panel; in that the fuel master cocks for each engine will be switched 'on', and the ignition switches for the engine being started are uncaged and switched 'on'. The order to start the engine is given by the pilot. On the engineer's panel, the start master switch is now 'on'. The buttons being held for the engine being started are the booster coil to 'on', the starter motor switch to 'on', and the priming switch has been set to the engine selected for starting and depressed in order to provide priming fuel to that engine. Once the engine fires and runs for the first few seconds, all indications are carefully monitored, checking the oil pressure and any signs of a fire warning. A fire warning light or lack of oil pressure means the engine must be immediately shut down. *(Keith Wilson)*

BELOW Once engine No 3 is running, engine No 4 is selected and started, followed in turn by Nos 1 and 2. *(Keith Wilson)*

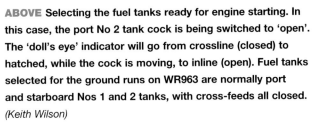

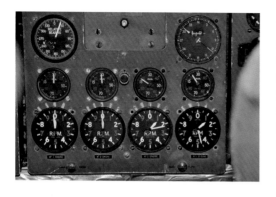

LEFT The pilot's centre panel includes the engine instruments and here it shows No 3 engine and No 4 engine are running. No 3 is at 'idle', and will now have its generator switched 'on', allowing the aircraft to generate its own power rather than relying on internal batteries or the ground power unit. No 4 has just started and will be pulled back to 'idle' – the momentary increase in revs being as the throttles are set at 1in open, but until the engine is running an accurate rpm cannot be set. *(Keith Wilson)*

LEFT Another view of the pilot's centre panel, this time indicating all four engines are now running. All four generators will have been switched to 'on' and the aircraft's inverters will now have been switched 'on', bringing to life some of the flight instruments. The priming switch is 'off', the start master switch is 'off', and now it's a matter of waiting for the engines to reach operating temperature before they can be exercised. *(Keith Wilson)*

set the throttle back to 1in open in order to get it running after it bursts into life on the priming fuel, only to run down again!

As the engine starts to run, the starter switch can be released. Then the boost coil can be released when the engine reaches around 400rpm, as the engine starts to pick up speed. As it climbs further and starts to smooth out, the priming switch can be released – being pushed again if the engine starts to falter. Finally all 12 cylinders kick in, the rpm gauge spins around to approximately 1,200rpm, and the engine is running.

Immediately you check to see that there is sufficient oil pressure on the gauge, and no sign of a fire warning light on the emergency panel above the main instrument panel. A lack of oil pressure or a fire warning means an immediate shutdown of the engine!

If all is well, the engine is settled at 'idle' – around 1,200rpm – and the DC generator is switched on for that engine, by holding the switch to 'reset' until the voltage comes up and stabilises at around 28 volts, before flicking it smartly to 'on' to bring it online.

The bomb doors are set to 'closed' to test the hydraulic system is functioning correctly and once the doors are closed, the next engine (No 4) can be started using the same sequence. Engines are started in the order 3, 4, 1 and 2.

As each DC generator is brought online, the

LEFT The engines are all started and are working up towards operating temperature. The variance in temperatures is reflected by the order of starting – Nos 3, 4, 2 and 1 – and you can see the temperatures are high to low in that order. The gauge for the No 1 charge temperature is stuck, and a gentle tap awoke it. The switches selected are charge temperature control to 'cold' with radiator shutters to 'manual' and 'open'. Normally, to achieve a faster warm up, you would use 'automatic', which would see them closed, and a thermostat-controlled system operates a pneumatic inching control to gradually open the shutter to maintain the correct coolant temperature. On WR963, the manual control to override is employed as the pneumatic system which supplies the shutters is also required for brakes, and during the ground running the brake pressure takes precedence, even if this means a longer warm-up. *(Keith Wilson)*

generator failure light goes from flashing yellow to solid yellow, before finally going out. The Nos 1, 2 and 3 inverters can now be brought online to activate various flight instruments, before the start master switch is switched 'off' and we now wait for the engines to warm up to operating temperature.

Exercising the props

Once the engines are warm (around 90–105°C), the propellers are exercised individually by opening the throttle to attain 30in of boost and around 1,800–2,000rpm; then the propeller control (pitch lever) is brought slowly down from 'max revs' to 'min revs'. This moves the propeller from its fine to coarse pitch and exercises the translation unit between the propellers.

The revs fall to around 1,550rpm and then the propeller control is moved back to 'Max Revs'; this exercise is repeated three times for each engine and propeller combination and is completed in the same order as the engine start.

The shutdown

Once the ground running is complete, the engines are all throttled back to their 1,200rpm 'idle' setting and allowed to cool to a temperature of 105°C maximum. The flaps are selected 'up' and the Nos 1, 2 and 3 inverters are all turned 'off'. The DC generators on Nos 1, 2 and 3 engines are switched offline, and on the pilot's command the slow-running cut-off switches for those engines are held down simultaneously until the engines shut down and the propellers come to a stop.

This leaves No 4 engine still running, and the bomb doors are now set to 'open', once again testing the hydraulic system. The Shackleton has no hydraulic pressure gauge so the only test of both pumps is by opening the bomb doors with No 3 running when starting, and closing them with only No 4 running.

The DC generator is switched 'off', and on the pilot's command, the slow running cut-off switch is held down, stopping the No 4 engine. When the propeller has finally stopped turning, all four ignition switches are set to 'off' and caged, the fuel master cocks 'closed', and the fuel tank cocks 'closed'.

The internal batteries are switched 'off', and all leads and hoses stowed. The rear door is

now opened and the steps positioned so we can disembark. For safety reasons, everyone must avoiding walking around the propellers while the engines are still hot. Then there is usually a short debrief to discuss any snags before we retire and let the aircraft cool down for a period.

Once cooled, the blanks can be placed back in their positions and WR963 sleeps again for *another* month . . . and we can leave happy in the knowledge that one of the last of a small handful of Shackletons in the world capable of doing so is still growling on defiantly.

The aircraft can (and has) been readied to a state where we can board and start it inside a few minutes – much as was done on QRA and is found inside the AEW.2's Flight Reference Cards. We don't do this often though!

The procedures used on WR963 vary slightly from those used by front-line Shackleton crews over the years, but it is done purely to prolong the life of the engines and the various systems in the aircraft. The Flight Reference Cards used throughout each run-up are derived from the original AEW.2 and MR.2 cards that were used in squadron service.

LEFT With engine Nos 1, 2 and 4 set to 'idle', engine No 3 is run up to almost 2,000rpm and the hydraulic propeller pitch adjustment is exercised. All four engines are systematically exercised in turn, giving the viewing public something spectacular to both watch and listen to – the Griffon roar! *(Keith Wilson)*

BELOW The most recent engine runs were conducted at Coventry on 21 March 2015, after the winter's maintenance programme and ahead of the regular spring and summer events. While all four engines are running beautifully, No 3 is going through its engine and propeller exercises. *(Keith Wilson)*

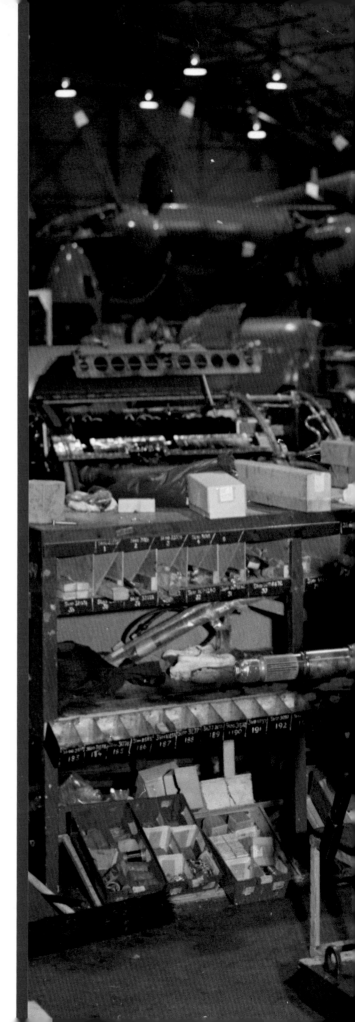

Chapter Seven

Maintaining the Shackleton

The Avro Shackleton was a large and complex aircraft, powered by four Rolls-Royce Griffon piston engines. Inside the fuselage was an extensive selection of communications equipment and radar, along with both offensive and defensive armament. Consequently, it required people with a broad range of engineering and technical skills to maintain it – including engine and propulsion ('grollies'), airframe ('riggers'), air electrical ('leccies'), navigational instruments ('nav leccies'), air radar, air radio and finally the armourers.

OPPOSITE A view of 8 Squadron's main maintenance hangar at RAF Lossiemouth on 16 December 1976. While work continues on a Shackleton AEW.2 in the background, three fitters are busy working on one of the Griffon powerplants in the adjacent engine bay. Other than the personnel, this view didn't change until the end of Shackleton operations in 1991. *(Crown Copyright/Air Historical Branch image TN-1-7635-4)*

ABOVE A carefully posed image of a Shackleton MR.2 of RAF Coastal Command undergoing routine maintenance at its base at RAF St Eval. The gantry being manoeuvred by the fitter in the foreground was rated with a safe working load of 2 tons, and could be used for lifting the engine powerplants. These gantries remained in Shackleton use right up to 1991, as did the coolant trolley. *(Crown Copyright/Air Historical Branch image T-320)*

RIGHT Shackleton MR.1A, WB823/T-N of 220 Squadron, was one of five aircraft from the squadron to have visited Ceylon during February–April 1952 and the aircraft is seen here undergoing routine maintenance during that trip. Members of the groundcrew had to be adaptable to be able to work effectively, as can be seen from the clever use of the wooden wheel chocks! Unfortunately, the propellers appear to have been refitted at least one spline out of position which, as well as looking wrong, could also lead to vibration issues. *(Crown Copyright/ Air Historical Branch image CFP-555)*

When despatched on deployments all over the world, the Shackleton's amazing carrying capacity allowed it to travel long distances while still transporting its own maintenance crew, with its ground support and spares on luggage panniers located in the bomb bay. As a consequence, the Shackleton earned a reputation for being self-sufficient and it was often despatched to far-flung locations over the course of its long and illustrious service career.

Enter Druid Petrie

Druid Petrie is the newest trustee with the Shackleton Preservation Trust and is the team's chief engineer. With his wealth of aviation experience, Druid is a key part of the team of volunteers trying to return WR963 back to the air. Nine years with 8 Squadron at RAF Lossiemouth on the Shackleton AEW.2 makes him the ideal man for the job.

Druid joined the RAF in 1978 and following the standard route of square-bashing at RAF Swinderby, he took on nine months of technical training at No 1 School of Technical Training (SoTT) at RAF Halton. His first posting was to RAF Valley. This was not part of his grand plan, however, as he had applied for Hong Kong, Singapore and some other far more 'exciting' places. The RAF thought differently. In the end, it was a good posting, both personally and professionally. From the professional point of view, RAF Valley was a training unit in Support Command. In his own words:

There was time to do everything by the book, and we had time to learn the book, especially the Air Publication 100B, which at the time was the bible of engineering in the RAF. The idea was, that if you knew this book inside out, you could be moved to anywhere in the world, and you would know what the engineering rules were for any form of aircraft engineering.

Druid's second posting was to Strike Command, with 8 Squadron, operating the Shackleton AEW.2 at RAF Lossiemouth. 'This was my favourite posting,' he said. 'The lifelong friends I made, the work I thoroughly enjoyed,

LEFT Shackleton MR.2 (Phase 2), WL789/A of 224 Squadron, being refuelled at RAF Gibraltar in March 1962. The refuelling truck was affectionately known as a 'hippo', and the type of ground power trolley behind the bomb bay doors can still be found in use today at many airfields. *(Crown Copyright/Air Historical Branch image PRB-1-22431)*

RIGHT RAF groundcrew at work topping up the port 26-gallon water/methanol tank on a 203 Squadron Shackleton MR.3 (Phase 2), WR987. The Griffon engines on the Shackleton were fitted with a water/methanol injection system which boosted the available power on take-off from 1,960hp to 2,435hp. This image was taken in June 1962 at RAF Kinloss during NATO Exercise Fairwind Seven. The same model of water/methanol trolley was used right up to the end of Shackleton operations, although the designation for the water/methanol had changed to AL24. *(Crown Copyright/Air Historical Branch image PRB-1-23204)*

RIGHT Shackleton MR.2 (Phase 3), WR748 of 205 Squadron. It is seen here undergoing maintenance to the propeller and hub on the No 1 engine at RAF Changi in 1968. The smaller round tool to the left of the working fitter is for removing and refitting the intershaft seal and nut assembly, and the large, cylindrical tool behind the fitter's right foot is the extractor for the rear propeller bearing, which is located behind the intershaft seal and nut. Great merriment was had when new engineering arrivals attempted to impress their peers, but failed to undo the intershaft seal and nut assembly. This was because it was furnished with a left-hand thread. *(Peter R. March)*

Life on the Shackleton AEW.2

The background and politics relating to the Shackleton AEW.2 entering service as a short-term fix to the UK's immediate and pressing AEW requirement is covered in Chapter 4. When the Shackleton AEW.2 was introduced, it was already an elderly aircraft, most of the airframes being at least 16 years old. In its favour, however, there were a large number of engineering and operating personnel familiar with the Shackleton, and its many idiosyncrasies, foibles, irritating habits and traps.

Druid recalls his time on the Shackleton with affection:

No 8 Squadron at RAF Lossiemouth was – for a number of reasons – a unique unit within the RAF. As well as being an 11 Group squadron based on an 18 Group station, 8 Squadron was the sole RAF operator of the Shackleton; in fact, it was the only front-line piston-powered squadron in the RAF. This fact alone meant that maintenance support was very different from the mainstream.

On most RAF stations, there were three sections to cater for aircraft maintenance requirements: the line, where the aircraft were sent off from and recovered to following flight; rectification, or 'rects', where all heavy repair work of an unscheduled nature was carried out; and scheduled, or 'sheds', where, as the name suggests, the planned or scheduled maintenance inspections were carried out. No 8 Squadron only had the line and the scheduled teams as the line also carried out the 'rects'. This meant that uniquely, 8 Squadron 'linies' became very knowledgeable about the inner workings and details of their charges, and very experienced at major rectification work.

The Scheduled Servicing Team who resided in hangar K-16 continued to produce superb results every time they released an aircraft from their care. Their attention to detail, their own devotion to duty and pride in their work was uniquely responsible in keeping these fine 'old grey ladies' in exceptional condition. Also residing in K-16 was the BAe CWP (civilian working party), who for many years tackled some challenging and difficult structural repairs,

ABOVE A view of maintenance work being carried out on an 8 Squadron Shackleton AEW.2 at RAF Lossiemouth on 16 December 1976. The later Griffon 58 engines and propellers were mounted in universal powerplant (UPP) assemblies, complete with their radiators and oil cooler. These could be installed in any one of the four positions without change or restriction. This reduced the total number of spare engines required to be held. *(Crown Copyright/ Air Historical Branch image TN-1-7635-1)*

and the Motor Club, which formed a large part of my life in Scotland.'

Next stop was RAF Woodhall Spa, famous for its role during the Second World War as the base for 617 Dambuster Squadron. He recalled, 'This was a temporary holding posting, in the Tornado engine bay, but I still enjoyed my time here.'

His fourth and final posting was strictly by invitation.

I was delighted to be accepted for service with the RAF Battle of Britain Memorial Flight at RAF Coningsby. After the hard work over the previous 13 years, I thought it was an exceptional reward. However, unlike some, I never lost sight that it was all about the aircraft, not the staff. This ended up being my final posting, and I consider it to have been the best possible way for me to end my RAF career.

After leaving the RAF in 1996, Druid continued to work in aviation maintenance and set up his own aviation business in North Wales in 2001, which he has run since. There he has 'taken great pleasure in restoring a number of written-off aircraft back to the skies while saving a couple of historic types from the scrapman'. Interestingly, Druid was part of the team of civilian contractors that reassembled the BBMF Lancaster following the replacement of its main spar back in 1995, making it the first and only Lancaster aircraft ever to be re-sparred.

saving many a Shackleton from early retirement. The men of the CWP were unsung heroes, and their contribution cannot be overstressed.

The 8 Squadron line was an open area apron of heavily reinforced concrete that had at one time been a clean, white colour. Years of Shackleton parking had changed that, each kite having successfully marked its territory. Each engine oil tank contained 32½ gallons of thick oil that turned black due to the high carbon content scoured from the engine. This black permanent marker trashed anything fabric based. Those engineers entrusted with maintaining the engines were termed 'grollies', and their old overalls were regularly washed in buckets of Avgas, as this seemed to be the only product that removed the worst of the blackened oil residue. Oil-soaked hair was generally avoided by the wearing of hats, but sometimes even hats couldn't cope. Oil-soaked hands had another drawback. The oil-lubricated locking wire and split pins guaranteed regular punctures of the skin, especially under the fingernails. Although one got used to this, it still hurt just as much as the first time – and still does years later!

'Grollies' were responsible for the engines and powerplant systems, including the control runs from the cockpit to the powerplants; propellers and their systems; fuel systems including tanks, pumps, valves and cocks; large sections of the pneumatic systems, including the pumps and engine associated pneumatic valves and rams; hydraulic pumps, AC generators, DC generators, and the external gearboxes that drove them.

The 'riggers' or airframe guys had their very own fun – with hydraulic fluid. Although the fluid used in the Shackleton wasn't as cacogenic as the high-temperature, high-pressure Skydrol (think VC-10, etc.), it still had an acrid smell, seeming to trap its aroma permanently in your nostrils and skin. It also left permanent red stains on your clothing, and was irritating to the skin, but far more so in the eyes.

The 'fairies', or air radio/air radar trades had to deal with slow warm-up times, heavy and large black boxes, and irradiating valves in the Second World War-vintage scopes. Coupled with this, they also had the joy and pleasure of playing with the world's largest, operable

airborne microwave oven, as the AN/APS 20 radar worked in the microwave frequency range. The 'nav insties' or navigational instrument trade were kept busy with maintaining equipment that most of them hadn't even seen in training. The 'leccies' or electrical

BELOW No 8 Shackleton AEW.2, WL756 *Mr Rusty*, undergoing deep maintenance at RAF Lossiemouth in February 1990. The large, ribbed square panel standing on end is the No 1 fuel tank access panel, installed under the wing between the inboard engine and the fuselage. The panel on the safety raiser to the right is the No 2 fuel tank access panel, installed on the upper surface of the wing between the engines. *(Keith Wilson)*

BOTTOM Another view of WL756 at RAF Lossiemouth, this time with the engine bay in the foreground. A powerplant is being assembled around a Griffon engine; the two radiators are already in position and the oil cooler is on the bench to the left. *(Keith Wilson)*

1 SPT volunteer Phil Woods on the starboard wing of WR963 at Coventry, attending to the inboard No 3 engine. The first task is to release the flap covering the oil filler, undo the Terry locking clip and then twist and release the oil filler cap. *(Keith Wilson)*

2 Carefully lift the dipstick to check on the existing oil level. For ground running only, the oil reservoir needs to contain a minimum of 19 gallons, but for flight, the normal level prior to flight is 32.5 gallons of oil per engine. Consumption of up to 1 gallon per engine per hour was considered acceptable, which was a much lower consumption than Bristol sleeve-valve radial engines. *(Keith Wilson)*

3 Carefully add the required quantity of W100 oil to the tank . . . *(Keith Wilson)*

4 . . . before rechecking the level. *(Keith Wilson)*

5 Access to the outboard engines is via a removable panel on the outside of the engine, as seen here on engine No 4. *(Keith Wilson)*

6 Volunteer Mark Ward is seen carefully adding the contents of a 1-quart container into the reservoir on the No 4 engine . . . *(Keith Wilson)*

7 . . . before checking that the contents are at the minimum required 19-gallon level. *(Keith Wilson)*

8 Thankfully, the Shackleton Preservation Trust has – in Sil-Mid Limited – a wonderful sponsor of lubricants for WR963. *(Keith Wilson)*

9 A fitter filling the No 2 engine inboard oil tank. *(Crown Copyright/Air Historical Branch image T-3040)*

10 The above sequence of images make an interesting comparison with how the process was carried out on the flight line at RAF Gibraltar in 1962. Here, Shackleton MR.3/3, XF702 of 206 Squadron, is being worked on by the local engineers utilising the David Brown tractor and the 400-gallon oil bowser trailer. This contained a more than sufficient supply of oil, with the hose able to deliver it into the oil tanks at speed – quite a contrast to the 1-quart containers used on WR963! The last operational Shackleton squadron – 8 Squadron – utilised a slightly more modern version of the David Brown tractor and a very similar 400-gallon Eagle oil bowser, the pump motor of the bowser being a Coventry Victor air-cooled petrol engine. *(Crown Copyright/Air Historical Branch image T-3042)*

ABOVE **Adjacent to the engine bay is the Griffon storage facility. In February 1990, seven engines were held awaiting installation, as and when required.** (Keith Wilson)

trade were kept busy maintaining aged wiring and tired generator systems as well as changing the occasional bulb.

And just to add a bit of spice to things, the Shackleton was a big aircraft, which meant that most things fitted to it were also big! Fully laden, she weighed 92,000lb without the bomb bay fuel tank. A complete mainwheel assembly weighed around 800lb while a Rolls-Royce Griffon powerplant complete with its propellers weighed around 2 tons; and there were four of those. Refuelling took a while. The eight fuel tanks were filled through six filling points with refuelling nozzles not much larger than a commercial diesel type, and the total capacity was 3,284 imperial gallons; around 10.5 tons of fuel alone.

Due to the effects of age, the Shackleton AEW.2 was a very labour-intensive aircraft. For each hour's flight, there would be in excess of 40 man hours of maintenance. For an aircraft capable of flying for over 14 hours, these figures don't bode well when the bean-counters are looking closely. But, at the time, the old girl was all we had. So we persevered, often interrupted, and not always supported.

Let me take you through a typical after-flight experience. The operating crew, who had previously borrowed the aircraft from the line after signing for it, now brought it back and signed it back in. Strangely, although they had taken it away fully serviceable, they always seemed to bring it back broken! A full, technical debrief is carried out, symptoms of all faults being accurately described by the operating crew, with the engineering crew making notes.

The first action is the after-flight servicing. Initially, the armourer would remove the bomb bay pyrotechnics; then he would then be redundant, potentially for many hours, as the last task on the aircraft would be reloading or fitment of the planned bomb bay load of flares and smoke floats. Then, like busy bees, members of every trade are whizzing around, on, in and under the aircraft, inspecting, resetting, tidying, and replenishing everything – oil, water/methanol, hydraulic fluid, coolant for the engines and the radar, water for the galley.

If there have been any reported major snags that may require prolonged engine runs, refuelling would wait until later. Then it would be time to fix the broken bits. A team of riggers would start jacking one side, or arc-lifting, in order to change a mainwheel because of a brake sac pneumatic leak, and investigating a leak from the flap hydraulic jack, and replacing a de-ice pump cell. The 'fairies' are swapping out a pair of APX 7 units (which were probably the most unreliable kit they had), a low-voltage power pack and a scope, because of failures. The 'leccies' are changing a DC generator carbon pile regulator because that generator is not load sharing, and a landing light filament (think bulb) – all 350 watts of it – because it has blown. The 'nav insties' are replacing a pilot's compass repeater because it is lagging behind excessively; and the radio altimeter control unit because the aural warning isn't working. Meanwhile, the 'grollies' are pulling the front propeller and translation unit off one of the engines because of a leaking intershaft seal; replacing one of the magnetos on another engine because of a dead mag'; pulling off an oil cooler on another engine to gain access to a failed starter motor; and ripping the automatic boost controller off the fourth engine because of a hang-up of the throttle when the pilot closed the throttles just before touchdown, leading to a swing on landing – never a good thing. They have also disconnected pneumatic pipework for a five-way valve replacement because the radiator shutters won't close fully on one engine, dropping the early warning filters for the routine 25-hour check, and preparing a new fuel flowmeter to replace the unit on one engine that failed in flight. And then they are removing the ducting for an inboard DC generator in preparation for pulling the generator (weighing around 80lb) in order to replace the quill drive 'O' ring because of leakage of the gearbox oil. An engine change has just been added to the list of work because of metal deposits in one early warning filter. And the aircraft will need a full defuel because the riggers have found a crack in the starboard inboard forward lower panel in front of the No 1 fuel tank.

Now, try to imagine all this happening outside on an open and exposed concrete apron . . . at night . . . in January in the north of Scotland . . . in a winter storm. At well below freezing – and with a wind chill of -10º, with snow in your eyes and ears and sheet ice on the ground. Think of the fun walking out on the frozen wings to do the refuel, probably in excess of 2,000 gallons. How long does it take to put 20 gallons in your car?

Remember the riggers and their cracked lower skin? This job requires the aircraft being placed into the K-17 hangar following the full defuelling, jacking the aircraft off the ground with trestles placed at the wing outboard trestle points. The weight of the inboard No 3 engine is supported on the front propeller with the propeller lifting hoist – SWL of 2 tons – and a team of guys, all trades, with screwdrivers in hand to remove the No 1 fuel tank stressed access panel, probably around 400 screws. All by hand – no electrical screwdrivers allowed here! Then the fuel tank is removed to gain access for the riggers to make up and fit a butt strap repair to the cracked forward panel.

It is 4 hours later, and the engine change has also been completed. Now, it is a simple job of putting it all back together, removing the aircraft from the jacks, dragging it back outside for the refuel and leak check. Then, a long night of ground runs for the engine change and the various engine defects that have had replacements; the AC generator ground runs for the 'fairies'; and the DC generator runs for the 'leccies'. And when all this is done, the aircraft will need a compass swing.

Remember the 'nav insties' had to replace a pilot's compass repeater. . . . A good swing will take 2 hours; a bad one will take two days. All of this must be accomplished before the aircraft can be declared 'serviceable' and ready for flight. Well, not quite; the ever-patient armourer can now load his stores into the bomb bay. Only then is she is ready to fly.

Why did we do it? Well, it was at the height of the Cold War and we had to react to the continual attempts by the Soviets to enter our airspace. The Soviets kept coming with their *Bears* and their *Bison*, and our very real mission was to keep them out. This is exactly what the AEW.2 version of the Shackleton had been produced for. Our venerable Shackletons were regularly required to 'scramble' and our contribution was to place an airborne radar platform in the air, at no notice, whenever it was required, 24 hours a day, 365 days of the year. Our Shack would control and vector our Lightnings and Phantom air defence aircraft on to the 'visitors'. This was called a Quick Reaction Alert (QRA), and we had to turn our broken aircraft around as quickly as possible in order to meet our QRA commitment – all the way up until 1989 when the Berlin Wall came down following the eventual collapse of the Soviet Union.

We rose to the challenge with pride and professionalism. Often, in order to get the aircraft back on line as quickly as possible, guile, ingenuity, initiative and imagination would be used. While on detachment to RAF Leuchars and due to a 'work to rule' by civilian crane operators, I used a Royal Engineers JCB to remove and fit propellers – not once but twice. Thank you RE Staff Sergeant Stewart.

When one of our aircraft broke at the sleepy base of RAF Topcliffe in Yorkshire, where there was no on-site support, I broke a number of standard RAF rules in this very non-standard situation. This included using a large aircraft tug I didn't have a licence to drive to open and close seized hangar doors and driving a truck I didn't have a licence for . . . and a few other minor divergences. I couldn't have been that bad a chap as my senior engineering officer (SEngO) backed me up completely, although I didn't find out for years just how much flak he took on my behalf.

On detachments to foreign lands, interesting things have happened. I once had to pull a Shackleton out of a collapsed drain, thankfully without ripping the tailwheel assembly off. However, a team putting WL756 into K-17 hangar at Lossiemouth weren't so lucky, the consequential repairs taking a good few weeks.

One aircraft on detachment to Norway ended up having a triple engine change, one for overheating, and two for running at maximum rpm for far more than the allowed time. On one return from a Gibraltar trip, we were inside the aircraft waiting for customs while one of the riggers started his after-flight inspection. He stopped, started giggling and laughing so hard he was curled up on the apron, unable to speak or breathe, but pointing up at the underside of the starboard wing. Ignoring our directive to remain on board the aircraft, I had to see for myself what tickled the rigger. It was the 8ft crack from the leading edge to the trailing edge! I once found the same problem on a serviceable aircraft that had been put into K-17. I guess you can miss spotting an 8ft crack, can't you?

On one detachment to Keflavik in Iceland, we had to have our Shackleton towed inside to allow me to carry out various repairs, the weather outside being too severe. It was rather difficult to concentrate on removing the broken bits while the USAF servicemen were in the middle of a TACEVAL [tactical evaluation] . . . inside the hangar. Blank rounds are very, very noisy – especially inside a large building.

Ground running of the engines was always a thing of pleasure. The noise a Shackleton made was both unique and unsilenced, and we made sure the locals – either in Lossiemouth or wherever we were in the world – knew we were still working. Off shift, I could hear the engine runs from my home over 8 miles away.

Following any engine rectification work, full functional and leak checks would be required. If the 'fairies' had an AC supply problem, we could be required to run the port engines in order to power the AC generators that supplied the radar in the air. If the 'leccies' had disturbed the DC generator systems in any way, the engine runs could be long and laborious. On these occasions, the two-man team inside running the engines would dial the ADF to 1053 and listen to Radio 1 on the medium wave band. And if the engine runs happened after 19.00 hours, we would dial up 1440 and listen to Radio Luxembourg. The man operating the flight engineer's panel would have to plug into the navigator's station to receive the ADF signal.

However, the poor sod acting as outside man only had a deafening intercom to hear the inside crew. Like the Shackleton, both these services have ceased operation.

Sometimes, unplanned events would happen during engine runs. I have experienced a spinner coming off at high power, blown a spark plug out, suffered an intake fire, had many exhaust fires, burst an oil cooler, rpm indication die, had a failure to shut down on more than one occasion, the failure of an automatic boost controller (ABC) allowing 81in of boost without water/methanol injection, and a whole host of other issues that help keep the mind focused on the task.

When seated in the pilot's seat, one had almost 10,000hp at your fingertips, and I only tried all of it once. The tailwheel moved 3ft sideways. Running the engines at high power on the ground entailed grabbing the yoke and pulling it into your chest and holding it with your right arm, standing with maximum effort on the brakes, and operating the throttles with your left hand in a controlled fashion. If there is a tail- or crosswind, the bucking and fighting of the control yoke leaves bruising of the ribs and inner arm. In the air, she is such a graceful and gentle girl, responding sweetly to each light input. It is obvious that her natural environment is not on the ground.

The Rolls-Royce Griffon ran before the Merlin, and served in the front line far longer, but for some strange reason doesn't enjoy the fame it truly deserves. In service with the Shackleton AEW.2, the Rolls-Royce Griffon 58 was an extremely reliable engine. It tended to be let down by the bits that were bolted on to it, such as magnetos, constant speed units, automatic boost controllers, starter motors, oil coolers, pneumatic rams and valves, ignition harnesses, ignition booster boxes, etc.

The continual delays with the AEW replacement programme meant the old girls had to soldier on far longer than anyone had planned for – and this led to significant supply and procurement problems. In the latter half of the 1980s, RAF Stafford, probably the RAF's largest storage facility, informed 8 Squadron that they either had to take on the total spares holding for the Shackleton, or it would have to be scrapped as RAF Stafford needed the room for more recent aircraft types in service, such as the Tornado. Consequently, the K-17 hangar ended up with a phenomenal collection of fuselage sections, bomb doors, wing control surfaces, tail sections, boxes of 'bits and bobs' . . . and a selection of virgin spar bars. These bars were all square section, very long and heavy, and all stamped 'Ultrasonically Tested'.

A further consequence of the delay was that no new contracts had been raised for the repair and overhaul of many of our consumable items. Three that come immediately to mind were magnetos, brake sacs and heater fuel control units, although there were many others. The Electrical Servicing Bay at RAF Lossiemouth was responsible for carrying out servicing and rectification of our magnetos. Unfortunately, a situation arose where one individual, a junior NCO, had lost interest, and merely sent the magnetos back out with a fraudulent 'serviceable' label. This came to a head one night when I had diagnosed a faulty magneto, and replaced it with a 'serviceable' Electrical Servicing Bay-serviced item, and the fault was still present. So we replaced it again – and the fault was still there. By the time you replace something twice and there is no change, those below you understandably doubt your diagnosis. By the time you replace something four times, your seniors are quite understandably doubting your diagnosis and decision making. By the time you replace something six times, you are starting to seriously doubt yourself. When you replace the magneto seven times, and then everything is all right, your sense of vindication is not nearly as strong as your homicidal feeling towards the individual who has certified the previous six. Following this event, magnetos appeared from RAF stock, probably RAF Stafford. However, they were an earlier mark, Rotax NT 12Js instead of the usual NT 12Ks. Once I had persuaded the SEngO that the only change was a difference in internal timing of 6° instead of 2, and this would only be seen as a slightly larger mag drop on the exhaust side, he was more than happy to accept these for our use.

In the latter part of the 1980s, we had horrific problems with leaking mainwheel brake sacs, the items that inflate and then push the brake shoes outwards against the brake drums. Brake

sac leaks were a regular problem, occurring at the rate of one a week across the fleet. Overnight, we went to every single one failing; often on first test application after fitting a mainwheel. It was quickly established that the problem was in manufacturing, and it would take some time to remedy.

However, we were still required to meet our commitment. The squadron boss decided that both the engine starts and engine run-ups would be done on the apron prior to taxi, and completed with the chocks in and the brakes off. Once completed, the aircraft would taxi over the chocks. We curtailed this idea after the revving of the engines up to 2,600rpm to jump over the chocks led to a Shackleton slamming on the brakes – just in time – to prevent it carrying forward on to the grass in front. At the same time, the heavy rubber chocks sprung by the mainwheels and assisted by the high-power propwash always seem to aim directly at the aircraft's tailplanes. It was soon established that the production problem was due to the new guys following the book; the experienced guys having been pensioned off. They were dragged back, at consultancy rates no less, to show the young bloods how to do it. Well, not everything you need to know is written down.

For an aircraft whose operating environment is in the North Atlantic or North Sea, especially in winter, heating isn't a luxury, it is an absolute necessity. However, there were no heater fuel control units available; well, not serviceable ones anyway. The contract had never been renewed years before, and the unserviceable ones were piled up in the station stores. When

ABOVE At the opposite end of the maintenance hangar in February 1990 was WR963 *Ermintrude*. The equipment in front of the starboard wheel is a ground hydraulic rig, powered by a Coventry Victor petrol engine, probably older than the aircraft. *(Keith Wilson)*

1 This image shows the sections that make up a front spinner backplate. An outer ring is attached to an inner ring with three attachment plates. The whole assembly is precision balanced, and the numbering ensures it will be assembled correctly in the balanced format. *(Barry Wheeler)*

2 Positioned in front of the rear propeller is the translation unit, manufactured by Martin-Baker, better known for their ejection seats. The translation unit is the device that 'translates' the change of pitch from the front propeller, the master, to the rear propeller, the slave. The rear propeller rack nuts can be clearly seen. The triangular-shaped plate with three holes is the translation unit front plate, and the holes are where the rear extension of the front propeller rack bolts align. Their nuts are fitted on the rear face of the front plate. *(Barry Wheeler)*

3 This image shows the match numbering of the rear rack bolts and their associated rack nuts – in this case 2B – where they pass through the translation unit. The translation unit is similarly marked on its rear face. All parts of the Shackleton's propeller assembly are precision balanced, including the spinners, propeller pitch-change dome and the translation unit. There are three sets of rack nuts, one set for the rear propeller as pictured, marked 1B, 2B and 3B. A second set secures the extended rack bolts that protrude from the rear of the front propeller through the triangular plate on the translation unit, marked 1A, 2A and 3A. The third set is installed on the front face of the propeller pitch-change dome and are marked 1, 2 and 3. *(Barry Wheeler)*

4 The torque of the front propeller is seen here being set to 900lb/ft (+/– 100lb/ft) by the Shackleton Preservation Trust's chief engineer, Druid Petrie. The propeller pitch-change dome is pulled forward off the rack bolts in order to prevent the rack bolts being twisted, prior to tightening the propeller. *(Keith Wilson)*

5 Following the torque adjustment of the front propeller, the pitch-change dome is then rotated to the correct position mark, and slid over the rack bolts, ensuring the top-hat bushes – which are also position-marked – are first placed on to the rack bolts. *(Keith Wilson)*

6 The propeller pitch-change dome being refitted over the propeller blade rack bolts; so named as they adjusted the blade pitch angles by a rack-and-pinion method. As the pitch-change dome moves fore and aft, the rack bolts are drawn or pushed, changing the blade angles equally. The heads of the rack bolts are cross-drilled with four holes, ensuring at least

one will line up when the rack nuts are correctly torque-set. *(Keith Wilson)*

7 The pitch-change dome secured to the front rack bolts with the three rack nuts, which are split-pinned. *(Keith Wilson)*

8 The dome front plate – known as the 'junk head' is positioned, and the central nut is tightened. *(Keith Wilson)*

9 The 'junk head' is secured to the dome with a ring of bolts and nuts which are split-pinned, and the central nut is wire-locked. *(Keith Wilson)*

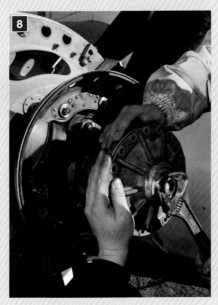

I did need to go and get one, the storeman was unable to help me at all as there were no serviceable heater fuel control units anywhere in the RAF. After mulling this over for a couple of seconds, I asked him if he could issue me with three unserviceable units. How to gobsmack a supplier? Ask for an unserviceable item from stock. Eventually, I got the three.

Locking myself in the tool store after briefing everyone that entrance was 'verboten', I stripped all three units. An hour later, I strolled up to the aircraft, installed a fuel control unit, tested it, and then presented myself in front of our SEngO. I openly told him what I had just done, in contravention of all standard engineering rules and procedures, and confirmed we had no manuals to refer to for what I had done. He said if I was comfortable with what I had done, and I considered it safe, he was happy to accept it. I 'fixed' a few more after that.

An unexpected occurrence happened one summer. While carrying out a high-power engine run, I obtained 52in on the boost gauge, giving around 2,500rpm. I then carried out a magneto performance check, turning off the left switch first. I received the most horrific electric shock as I moved the switch to off. After a quick curse, I moved the switch back on, throttled back and we shut the engine down. I dug out one of the 'leccies' and related my tale of woe with my fingers still tingling.

The 'leccie' couldn't find a problem, so I took him with me, and we started the engine again.

With 52in boost set, my bare left arm resting on the window sill while my left hand held the throttle, left switch off . . . the 'leccie' could hear my cursing above the open-exhausted 37-litre giant outside the window. Switch back on. I got him to try the switch. He didn't get a shock. I realised he wasn't holding anything else while he tried the switch, so I tried again, but without my arm resting on the window sill. No shock. He tried it while holding the co-pilot's grab handle . . . there was a noticeable Celtic edge to his curse. It turned out that there had been a batch of switches that had been produced, and the contact faces had been 'anodised' for corrosion protection. The problem was, anodising is not a good electrical conductor, but my bare arm was, as the 200-plus volts back-EMF testified. Old stock switches were 'obtained' from the MR.3s at Cosford, and even the T.4 at Strathallan, the cockpit of which now resides beside WR963 at Coventry.

Maintaining the Griffon engine

The powerplants were designed for ease of replacement. Predominantly all disconnection points were accessible. However, there were two items that weren't 'Murphy-proof': the manifold air pressure (MAP) pipe and the fuel priming pipe. They were the same to look at and both had a standard right-hand thread connection. The clue that there was a problem was on initial start following a powerplant change. As the chap on the engineer's panel hit the primer, the MAP gauge would rise rapidly.

Most items had disconnection points on the fireproof bulkhead. These included electrical, pneumatic, oil, fuel and breather. The engine controls, throttle and propeller were disconnected at the engine. The powerplant was attached to the aircraft at four points. The method was a common one and comprised a sleeve with a slot cut through one side, and tapered internally. Into this was fitted a tapered bush. The further the tapered bush was inserted, the further the slotted sleeve tried to expand, making a tight fit between the powerplant and the engine. To hold the tapered bush in place, a bolt with a spreader

cap was installed through the centre of the hollow tapered bush, secured with a castle nut and split pin. The spreader cap prevented the bolt from withdrawing out of the tapered bush, ensuring the attachment point remained rigid.

The second 'Murphy' was fitting the spreader cap to the wrong end, thereby encouraging the taper pin to withdraw, which was actually the way to initially loosen them when removing the powerplant. A slide hammer was then used to withdraw the slotted sleeves. The top attachments were removed first, and the powerplant was then slowly tipped forward, pivoting around the two lower attachments, to allow the main oil connections to be fully unwound – these large-diameter hoses always seemed to be about 2in too long. Blanks (rags used mostly) were then installed into the bulkhead connections and the oil pipes, and then the powerplant was lifted back up until the top attachments were aligned. The lower attachments were then undone. The powerplant lifting sling incorporated a screw adjustment for the lifting fulcrum, permitting the centre of lift to be adjusted fore and aft, thereby allowing a powerplant to be lifted with or without propellers. A complete propeller assembly weighed around 600lb.

The powerplant lifting sling connected to the powerplant at four points. These were raised webs on the engine rocker covers. In the middle of each web was a small-diameter hole, similar in diameter to a pencil. Four small pins were inserted into these little holes, and the whole powerplant could be lifted and supported from these. Each little pin was lifting around 1,000lb. Of course, it was perfectly safe. It was designed by Rolls-Royce . . . but nobody ever stood under a suspended powerplant.

To facilitate installation of a powerplant, two assembly 'bullets' were used. These helped guide the slotted sleeves into place, aligning any attachment points that, due to tolerances or just distortion, didn't align perfectly.

Powerplant replacements were a regular occurrence, normally due to an engine reaching its life limit. Engine failures were, in all honesty, a rarity, but did happen. A difficult one could take over 8 hours from start to finish; bad ones even longer. I seem to remember the best I did was – from panels off to signing up the job at the end

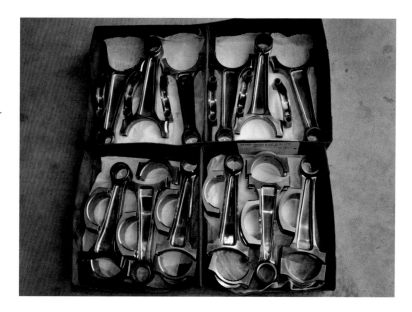

ABOVE A full set of conrods from a Rolls-Royce Griffon 58 engine awaiting NDT (non-destructive testing) as part of a major overhaul at Retro Track and Air. Rolls-Royce manufactured these conrods in this clean and polished condition. The upper set are known as 'knife' rods and the lower set are known as 'fork' rods. The fork rods are installed on the crankshaft big end journals, and they also act as the big end journals for the knife rods. This configuration means the two cylinder banks are not offset or staggered, as is the case when conrods on 'vee' engines are paired on to one crankshaft journal. *(Keith Wilson)*

BELOW A crankshaft from a Rolls-Royce Griffon engine in a stand at Retro Track and Air. The next stage of its overhaul process required a visit to the Magnaflux metal particle inspection machine used to check the crankshaft for any signs of fatigue cracking. The crankshaft is sprayed with a special green fluorescent, magnetic ink and once the white light is extinguished the object is subjected to ultraviolet light, under which conditions any signs of fatigue cracking become visible. *(Keith Wilson)*

ABOVE One of the Retro Track and Air technicians is seen working on a Griffon camshaft-and-rocker mechanism (CRM) assembly. Like the conrods, these were finished to the highest standard during production. *(Keith Wilson)*

after ground runs – just under 4 hours.

Each powerplant had an oil supply of 32.5 gallons of thick, black oil. This had a tendency of staining everything it touched. If an engine had suffered a conrod failure through the crankcase, it always seemed that most of the oil tank contents ended up around and under the engine . . . which then ended up on the floor of the apron, hangar, taxiway, etc.

Although powerplant replacements were carried out in the field, there always seemed to be a crane nearby. This was the preferred option. The aircraft has three hard points on the top surface of the wing behind each engine. These are to allow the attachment of a frame to permit replacement of a powerplant where a crane isn't available. The same three hard points could also be used for a frame to facilitate replacement of the external gearboxes mounted behind the fireproof bulkhead, and to allow

RIGHT The Rolls-Royce Griffon 58 supercharger assembly undergoing overhaul at Retro Track and Air. There is no intercooler – as fitted to 60-, 70- and 80-series engines – as the 50-series were low-altitude engines only. *(Keith Wilson)*

replacement of the rather heavy DC generators mounted on to the external gearboxes. Although the chaps doing the scheduled servicing in the confines of the warm hangar used these frames, we tended to use brute force and ignorance, along with lashing tape as a sling.

When carrying out powerplant replacements away from base, it was always a concerning time when the crane arrived. We didn't know any of these strangers, and didn't know how good an operator they were. Our fingers were in close proximity to crush areas when undoing the main oil pipes, but more especially when trying to disconnect and reconnect the four main bearer attachments. On more than one occasion, a crane operator's parentage was questioned repeatedly and, once or twice, a different operator had to be found to replace the jerky fool who couldn't smoothly control his crane.

Of course, when carrying out some of these unscheduled powerplant replacements, the senior engineering officers of our host station would show a keen interest. This was normally replaced by shock and horror, especially when they saw the extent of territory marked by the smiling Shack!

Magneto problems

The Griffon went against aviation convention. In order to ensure a level of safe redundancy, all post-First World War piston aero engines have predominantly employed two separate and independent ignition systems. Rolls-Royce was convinced they could design a safe system utilising a single body incorporating two ignition systems. Of course, the concern with this is twofold. Firstly, there is only one, 'shared' drive to the body. Secondly, a catastrophic failure of one half of the magneto could have rather detrimental consequences for the other. To my knowledge, there was never a recorded failure of the drive system. During my time working on Griffon engines – 8½ years on Shackletons, followed by 4 years on Griffon-engined Spitfire PR.XIXs, I saw a number of severely trashed dual magnetos. The most common failure was 'creep' of the top rotor arm machining away at the brass contacts around the circumference of the top rail, followed by eventual detachment of the top rotor arm – a reasonably large piece of

1 The forward edge of the bomb bay on Shackleton WR963 features a number of cable and actuator rods. The main white bar just below the centre is the connecting rod between the two pilots' flying control yokes. The chains, rods and cables are for the four sets of engine controls, connected to both pilots' pedestals. Being located in the bomb bay, these suffer with exposure to the elements and require regular cleaning and lubricating.
2 SPT volunteer Mario McLaughlin moves carefully along each cable and control run spraying with a WD-40 aerosol.
3 SPT volunteers Mark Ward (left) and Pete Buckingham start the intricate cleaning process on all the cable and control runs on WR963. *(Keith Wilson)*

brass. Needless to say, all these bits and pieces of brass and insulation material have to go somewhere, and that is down. On the bottom of the magneto body, Rotax (the now-defunct British firm that specialised in superb electromechanical products for the aviation industry, not the foreign manufacturer of small engines for the leisure industry) placed both sets of condensers and points. The potential for disaster appeared a certainty. However, on every occasion this type of failure happened, the only comment from the crew was 'during the after-landing checks, on doing the magneto checks, we noticed that the exhaust side was dead, no rough running, in flight or on the ground'.

Lycoming, the well-known American engine manufacturer, has employed dual magnetos on some engine models since the 1970s, but Rolls-Royce beat them by decades.

Getting WR963 back into the air

Some 24 years ago, back in 1991, the last Shackletons flying – 8 Squadron's AEW.2s – were retired from service with the Royal Air

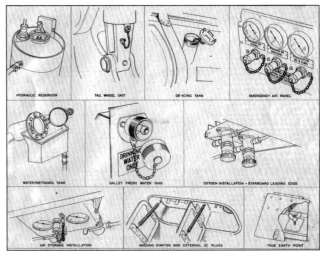

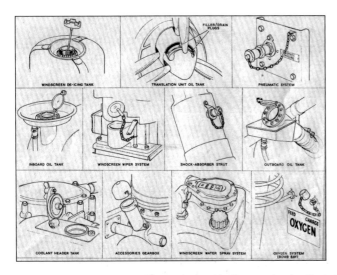

Force and sold off at auction by Sotheby's. The South African Air Force, the only other operator of the Shackleton, had retired their MR.3s previously. In total, 185 Shackletons were built, and today the Shackleton is quite well represented with around 10% still in existence, either as complete aircraft, semi-complete airframes or major sections. These are in museums and privately owned collections in the UK, South Africa and one in the USA. None are airworthy.

But that is about to change. In 2012, the Chairman of the Shackleton Preservation Trust (SPT), owners of Shackleton AEW.2, WR963, announced the very real intention to get '63 flying again.

Three things are required to return an aircraft to airworthy status: aircraft paperwork and

documentation, facilities and money. Following the transfer from British Aerospace in 1993, SPT own the design rights to all Shackleton aircraft. They also have WR963's technical records and history. Coupled with this is their substantial holding of serviceable parts and spares. Initial enquiries with the UK Civil Aviation Authority highlighted the CAA's interest in the project, but the bureaucracy and financial sums involved were eye-wateringly unjustifiable. The Federal Aviation Administration (FAA) in the USA was approached and, as well as being fully supportive to the project, their administration requirements and costs were considered appropriate to the task ahead. In December 2014, the FAA gave the green light to put WR963 back into the air.

Now, the hard work starts, and over 60 years after '63 first flew, a new chapter in the story of the Shackleton begins.

Would I do it all again?

The Shackleton AEW.2 – the 'short-term solution' – enjoyed a career that lasted as long as the Maritime Reconnaissance Shackleton MR.2. This venerable and gracious queen of the skies gave a total service of 40 years. I was involved in a small part of this – a mere 8½ years – but it formed a large part of my life, and the same was true for many others. We worked hard, but we played hard too. I enjoyed almost every minute of it, and I especially enjoyed the pleasure of working with some truly exceptional people.

Would I do it all again? Damn right I would.

ABOVE The Shackleton Preservation Trust has a secure facility located offsite where it stores its vast collection of Shackleton spare parts. Seen here is just a very small part of that inventory. To the front is a spare luggage pannier, while all of the surrounding boxes also contain spare parts, many of which may be required to get WR963 into the air again. *(Keith Wilson)*

LEFT Spare ailerons and elevators stored in racks against the wall. *(Keith Wilson)*

LEFT No longer in production, the SPT has around half-a-dozen spare mainwheel assemblies and tyres. The ridged items to the right side are new brake drums. *(Keith Wilson)*

BELOW Two of six mid-life Rolls-Royce Griffon engines held as spares for WR963. *(Keith Wilson)*

Chapter Eight

Shackleton survivors

Sadly, at the time of writing, no Shackleton is currently flying anywhere in the world. However, the small but dedicated Shackleton Preservation Trust at Coventry are determined to see WR963 return to the skies, although a considerable amount of work and a tidy sum of money will be required.

OPPOSITE With no airworthy Shackleton aircraft anywhere in the world, it is refreshing to know that the Shackleton Preservation Trust is actively trying to get their AEW.2/MR.2, WR963, back into the air. It regularly undertakes engine runs at its Coventry base. The sight and sound of four Griffon engines roaring is quite memorable! *(Keith Wilson)*

United Kingdom

Since the Shackleton Preservation Trust acquired WR963, they have managed to get all of the engines running and this they do on a regular basis. The sight and sound of four Griffon engines running is something to behold! For a small donation, you can gain access to the aircraft while this happens. The SPT have also managed to get the aircraft into a taxiable condition, providing visitors with an opportunity to witness a Shackleton moving under its own power at Coventry Airport a couple of times a year.

Interestingly, WR963 left RAF service as an AEW.2 following its time with 8 Squadron at RAF Lossiemouth and was acquired by the Shackleton Preservation Trust, founded by David Liddell-Grainger. The aircraft was flown to Coventry on 9 July 1991 with 15,483 hours on the airframe, accompanied by WL790, which had also been acquired by the Trust. It was intended that a Shackleton would be flown as a display aircraft at airshows and events and a large package of spares was also purchased to support this. Unfortunately, there was little co-operation from British Aerospace (who held the design rights to the Shackleton) and the Civil Aviation Authority (CAA). Consequently, the Shackletons were handed over to Air Atlantique in an attempt to find a way of getting one into the air in civilian ownership.

WL790 was prepared for flight and left for the USA in 1994, where it continued to fly for a further 14 years – a massive effort from all involved. Now, WL790 has now been prepared as a ground exhibit at the Pima Air & Space Museum in Tucson, Arizona, and is not expected to fly again.

Meanwhile, WR963 wasn't so lucky, becoming a source of spare parts for WL790. However, in 1997 the SPT was re-formed at Coventry to restore WR963, under the watchful eye of Squadron Leader (Retired) John Cubberley. WR963 has had the large radome removed from underneath the front fuselage and has been repainted as an MR.2, with the codes 'B-M' added.

Within a decade of its acquisition by the new Trust, WR963 was rolling down the runway at Coventry under its own power, almost at the point of flight. Once this landmark had been reached, a feasibility study commenced to get WR963 back into the skies. Apparently, WR963 has 594 flying hours remaining on the airframe and that represents a lot of airshow attendances.

In 2012, the CAA was approached, and answers found to the question of the viability of flying a Shackleton in the UK. Currently, the long-standing spar life issue remains the only thing keeping WR963 on the ground. However, with flight in mind, the registration G-SKTN was reserved in February 2013 and the fund-raising began.

More information regarding ground-running and the continued restoration of WR963 can be found on their website at http://www.avroshackleton.co.uk.

Museum exhibits

A total of six 'complete' Shackletons are displayed within museums in the UK. Without doubt, the best preserved of these

RIGHT Without doubt, the best-preserved museum Shackleton is the former 8 Squadron AEW.2, WR960, displayed at the Museum of Science and Industry (MOSI) in Manchester. The aircraft was delivered to the museum by road in January 1983, then reassembled and placed on display the following month. *(Museum of Science and Industry image 10672141)*

is the former 8 Squadron AEW.2, WR960, displayed at the Museum of Science and Industry (MOSI) in Manchester. WR960 was originally manufactured as an MR.2 and made its first flight on 5 February 1954. It served with a number of RAF squadrons before being returned to Hawker Siddeley at Bitteswell in May 1971 for conversion to AEW.2 standards. It was sent back to Lossiemouth in May 1972 for the installation of its 'new' radar. The aircraft was taken on charge by 8 Squadron the following day and the name *Dougal* added. It remained in service with 8 Squadron until November 1982 when it was flown to Cosford and allocated the maintenance serial 8772M. It had accumulated a total of 9,712 hours' flying time. It was later dismantled and prepared for transport by road for exhibition at the MOSI where it arrived in January 1983. It was reassembled and placed on display the following month, where it resides today although, sadly, it remains locked and the public are not permitted to view the interior of the aircraft.

On display with the Imperial War Museum (IWM) at Duxford is Shackleton MR.3/3, XF708/C. This aircraft was delivered direct to 201 Squadron in March 1959 and served with a number of squadrons before being flown to Kemble in January 1972 and placed into storage. On 23 August 1972, XF708 was flown to Duxford for exhibition with the IWM. It remained outside at Duxford and suffered over the years, but has recently been dismantled and moved to the Restoration Hangar where it will be slowly restored before being returned to public display.

The Newark Air Museum (NAM) at Winthorpe also have a Shackleton MR.3/3 on display – theirs being WR977/B. The aircraft was delivered to the RAF in August 1957 and transferred to 220 Squadron as 'L' the following

month. Like most Shackletons, WR977 served with a variety of squadrons before ending its days with 203 Squadron in August 1970. Here it remained until allocated to practice fire-fighting duties at Thorney Island, but the allotment was cancelled and, instead, the aircraft was flown to Finningley in November 1971 for display within the station museum after allocation of the maintenance serial 8186M. Its stay here was relatively short and in early 1977 the aircraft was dismantled and delivered by road to the NAM at Winthorpe where it arrived at the beginning of

BELOW Shackleton MR.3/3, XF708, was flown to Duxford on 23 August 1972 for exhibition with the IWM in 203 Squadron colours as 'C'. By October 2013, XF708 was showing the effects of more than 30 years outside at Duxford and it was moved inside the Restoration Hangar away from the elements. The outer wings were removed to get the aircraft into the limited available space and it was waiting the next stage of its rebuild. This is expected to take at least two years to complete. *(Keith Wilson)*

ABOVE Shackleton MR.3/3, WR977/B, was flown to Finningley in November 1971 for display within the station museum. Its stay was relatively short and in early 1977, the aircraft was dismantled and delivered by road to the Newark Air Museum at Winthorpe where it arrived at the beginning of May. After reassembly, the aircraft was placed on display where it currently resides, painted as 'B' of 42 Squadron. The impact of the elements is just beginning to become apparent in this image, taken on 16 April 2014. *(Keith Wilson)*

May. After reassembly, the aircraft was placed on display where it currently resides, painted as 'B' of 42 Squadron.

Interestingly, the Gatwick Aviation Museum (GAM) at Charlwood had two Shackleton MR.3/3s on display although, at the time of the author's visit, one of them was being dismantled for a road trip to a new owner at Bruntingthorpe. The aircraft remaining with the GAM is WR982/J, which was delivered to 23 MU in February 1958. It was issued to 206 Squadron the following month where it operated as 'B'. Like most Shackletons, WR982 has been

'around a bit' including a number of visits to the A&AEE at Boscombe Down for various Ministry of Aviation and manufacturer's trials work. It continued in service until flown to Cosford in September 1970 for use as an instructional airframe with No 2 SoTT where it was allocated the maintenance serial 8106M. In 1988, it was offered for sale by the Ministry of Defence in 'fair condition'. It was acquired by Peter Vallance and moved to Charlwood where it remains on display today. Happily, tours inside the aircraft are available, hosted by well-informed and enthusiastic former Shackleton aircrew.

The second Shackleton that was at Charlwood is another MR.3/3, WR974/K, which was on public display. In an effort to 'slim-down' the collection to appease local planning issues, it has been sold and is now at Bruntingthorpe. WR974 made its first flight on 1 May 1957. It did not enter squadron service but was released into the care of its manufacturer and the A&AEE for various trials work including tropical trials

RIGHT In 1988, WR982/J was offered for sale by the Ministry of Defence in 'fair condition'. It was acquired by Peter Vallance and moved to Charlwood where it remains on display with the Gatwick Aviation Museum. It is seen here on 27 September 2014. *(Keith Wilson)*

at Idris, Libya, in 1957. WR974 continued as a trials aircraft until being delivered to the ASWDU at Ballykelly in June 1966. Trials work continued until August 1968, when the aircraft was delivered to 203 Squadron. It remained in squadron service until December 1970 when it was flown to Cosford for use as an instructional airframe with No 2 SoTT, with the maintenance serial 8117M. It too was offered for sale by the MoD in 1988 and was also acquired by Peter Vallance who moved the aircraft to Charlwood where it remained on display as 'K'. At the time of writing, WR974 had completed its road trip and was being reassembled at Bruntingthorpe.

The final airframe in the UK is not quite complete. The fuselage of MR.3/3, WR971, is displayed with the Fenland and West Norfolk Aviation Preservation Society (FWNAPS) at their museum site outside Wisbech – in four pieces! The main fuselage section is displayed allowing visitors to step inside and view the various operator stations on board the aircraft. Nearby, the remaining three sections of the fuselage can be seen. WR971 made its first flight on 28 May 1956. It was retained by Avro for development work until December before being delivered to the A&AEE at Boscombe Down. Following the crash of WR970, the first MR.3, it was returned to Avro for further development work. After almost 5 years, WR971 was allotted to 206 Squadron but was actually delivered to 120 Squadron where it operated as 'E'. It continued in service with various squadrons until being flown to St Athan in December 1970, when it was transferred to No 2 SoTT at Cosford with the maintenance serial 8119M. Here it remained until offered for sale by the MoD in 1988. The fuselage was acquired for display at Narborough but was later moved by road to

the FWNAPS site adjacent to Bamber's Garden Centre in West Walton Highway, where the fuselage sections reside today.

Gate guardians

Only one Shackleton currently remains on display as a gate guardian and that may not be for much longer! At the former RAF Station St Mawgan, now renamed Newquay Cornwall Airport, Shackleton AEW.2, WL795/T, is displayed with the Inter-Service Survival School, which operates an RAF enclave at the airport. WL795 made its first flight as an

LEFT The second Shackleton at Charlwood is another former Cosford-based MR.3/3, WR974/K, which until recently was on public display with the Gatwick Aviation Museum. In an effort to streamline the collection, WR974 has been sold and was dismantled before relocating to Bruntingthorpe where it is currently being reassembled. *(Keith Wilson)*

BELOW The fuselage of Shackleton MR.3/3, WR971, was acquired for display at Narborough but was later moved by road to the FWNAPS site adjacent to Bamber's Garden Centre in West Walton Highway, where the fuselage sections reside today. The main fuselage section is open at either end, allowing visitors to step inside and view the various operator stations on board the aircraft. *(Keith Wilson)*

MR.2 on 17 August 1953 and was delivered to 23 MU in September. It remained in service with a variety of Shackleton squadrons before being returned to Hawker Siddeley at Bitteswell in February 1972 for conversion to AEW.2 standard. It was allocated to 8 Squadron and delivered to Lossiemouth for the fitting of the radar before joining the squadron at Kinloss in October 1972, where it was named *Rosalie*. It moved with the squadron to Lossiemouth in August 1973 and there it remained until withdrawn from service in 1981. It was allocated to crash rescue and fire practice duties at St Mawgan and flown there on 24 November. However, it was never used in that role and remained in open storage until 1988 when it was prepared for display on the station. It was returned to MR.2 configuration before being placed on display as 'T' – where it remains today – voluntarily maintained by the Cornwall Aviation Society. However, the RAF plans to

dispose of the airframe which has been put out to tender. At the time of writing its fate was unknown.

Cockpit sections

Two Shackleton cockpit sections remain in the UK, while a third is now in Holland.

VP293 is the oldest remaining Shackleton and has had an interesting history. It was first flown as an MR.1 on 18 July 1951. It was issued to 236 OCU in September where it operated as 'C-U'. It continued to serve with a number of squadrons including 224, 42 and 206 before being returned to Avro at Langar in August 1956 for conversion to T.4 configuration. After a short spell at the A&AEE, VP293 was delivered to the Maritime Operations Training Unit (MOTU) at Kinloss in March 1960 and coded 'X'. In January 1964, VP293 was purchased by the Ministry of Aviation (MoA) and transferred to the Royal

Aircraft Establishment (RAE) at Farnborough for trials work, where it continued to serve until disposed of in May 1975.

It was acquired by the Strathallan Collection and delivered there in January 1979 where it remained on display. Sadly, following the demise of the collection, VP293 was broken up on site. The nose section was acquired and preserved by Norman Thelwell, initially at the Norwich Aviation Museum, Horsham St Faith. VP293 has now been lent to the Shackleton Preservation Trust at Coventry where it is displayed, on a trailer, still carrying its MOTU code letter 'X'.

The cockpit section of T.4, WG511, is currently displayed with the Flambards Experience at Helston, Cornwall. WG511 started life as an MR.1A and made its first flight on 7 April 1952. It was issued into service and joined 42 Squadron in June where it operated as 'A-A'. It also served with 120 Squadron before being transferred to Avro at Langar for conversion to T.4 configuration in May 1956. Upon completion, the aircraft joined MOTU in August 1957 where it remained until withdrawn from service and eventually struck off charge (SoC) in August 1966. The nose section was converted to exhibition use by 71 MU at Bicester and it travelled the country from 1968 to 1971. In August 1974 it was donated to the RAF Museum and was later sold on to the Cornwall Aero Park – now the Flambards Experience – where it resides today.

The final cockpit section – of AEW.2 WL756 – having spent almost 8 years at Caernarfon Airport, is now with a collector in Holland. WL756 was first flown as an MR.2 on 1 April 1953 and delivered to the Overseas Ferry Unit at Benson for delivery to the Middle East Air Force (MEAF) at Luqa, Malta, where it operated with 37 Squadron as 'G'. Sadly, WL756 was involved in two minor incidents. The first took place in May 1954 when the pilot stalled the aircraft on final approach, resulting in the tailwheel assembly collapsing on touchdown; the aircraft was transferred to 137 MU at Safi for repairs the same day. After being returned to service, it was damaged again in November 1957 and was repaired by 103 MU at Akrotiri before resuming service.

In January 1971, WL756 was flown to Hawker Siddeley at Bitteswell for conversion

to AEW.2 configuration, making its first flight as such on 22 March 1972. It was flown to Kinloss in May 1972 and issued to 8 Squadron. After its 'new' radar had been fitted at nearby Lossiemouth and the aircraft named *Mr Rusty*, it was returned to Kinloss and entered squadron service. WL756 was withdrawn from service in 1988 having flown almost 13,700 hours. It was transferred to the Station Fire Dump at St Mawgan, where it remained for many years. Thankfully, the nose section of the aircraft was 'rescued' by former 8 Squadron Shackleton engineer (and enthusiast) Druid Petrie and was moved to Caernarfon Airport in February 2006.

ABOVE Sadly, with the demise of the Strathallan Collection, VP293 was broken up on site. The nose section was acquired and preserved by Norman Thelwell, initially at the Norwich Aviation Museum. The nose section of VP293 has now been lent to the Shackleton Preservation Trust at Coventry. *(Keith Wilson)*

BELOW The cockpit section is all that remains from Shackleton AEW.2, WL756. It was rescued by Druid Petrie from a scrapyard in 2004 and stored at Caernarfon Airport for many years with the 8 Squadron badge still visible on the nose. In 2014, Druid finally parted company with the section, and it now resides with a collector in Holland. The maintenance serial 9101M was also allocated to this airframe when it was withdrawn from active service in 1988. *(Druid Petrie)*

Druid Petrie takes up the story:

In around 2003/4, it was brought to my attention that the cockpit section of WL756 survived in Orchard's Scrap Yard at St Austell, or 'Zaastel', as the locals down there say. So I grabbed my wife and we travelled down to deepest Cornwall so I could have a look at an old friend. Mr Orchard Junior allowed me to go and have a look. The pitiful sight that confronted me was heart-wrenching. A once-proud aircraft reduced to a severed nose section, lying on its side on top of an ISO container, covered in moss and debris, and a .22 bullet strike that hadn't penetrated the nosegunner's bulletproof windscreen – so that had lived up to its name! She had been gas-axed following her days as a training aid for the fire crews at St Mawgan, and still bore the scars of her final duty for the Crown. What I thought to myself was, 'Poor old girl'. What came out of my mouth was 'How much?'

I agreed a four-figure sum, and made arrangements to collect some time in the future. On the trip home, my wife was fairly quiet until we were passing Bristol, where she enquired 'What possessed you to spend that much for a pile of scrap metal?' And for the first time in my life, I had to answer 'I'm sorry, I don't have a rational reason whatsoever.'

A number of months later, I drove down to Orchard's in my trusty, old, 300,000-mile Discovery and a borrowed car trailer to collect my 2-ton folly. The return trip was quite memorable. One chap was so engrossed in looking what was behind my car that he didn't realise the car in front of him had stopped. I decided that cowardice was the best decision and continued

driving. The trip up the M5 and M6 were interesting. On the day, there was a 50mph westerly blowing across the UK which made the trip even more interesting. However, it was very entertaining to note that we were a 50mph traffic jam as everyone overtaking slowed right down to have a good look at *Mr Rusty*. The motorway policeman decided it'd be 'too much paperwork' and after the obligatory slow-down-and-gawk, kept going, never to be seen again. The usual snarl-up at the M5/M6 junction was made bearable as everyone who was as equally stuck as I was, insisted in chatting about '56.

After a good wash, *Mr Rusty* resided outside my hangar for a couple of years. We then moved into slightly larger premises and she was moved inside for the first time in almost two decades. When work got busy, she was moved outside behind my hangar, where she remained until I sold her last year to a collector from Holland.

In reality, I was just the latest custodian of *Mr Rusty*. Of the whole 8 Squadron fleet, I had personally flown more hours in her, and had crew-chiefed her for the 40th anniversary bash at Woodford. So you could say she had got under my skin.

A number of poorly informed sources would have you believe she was part of the 'Caernarfon Airworld' collection, but nothing could be further from the truth.

Derelict airframes

At Long Marston aerodrome, south-west of Stratford-upon-Avon, the virtually complete but derelict airframe of MR.3/3, WR985/H, resides. It is a pitiful sight!

BELOW At the end of its service life WR985 was transferred to Cosford for use as an instructional airframe. In 1988 WR985 was offered for sale by the MoD and was sold to a private purchaser who moved the aircraft to Long Marston. Here it remains today, although it is in appalling condition and in immediate need of some TLC! *(Richard Vandervord)*

WR985 was issued to 206 Squadron in June 1958 where it adopted the code 'E'. The following year it went to the A&AEE at Boscombe Down in connection with trials for the Maritime Tactical Position Indicator. With trials completed, it was returned to squadron service until delivered to Avro at Langar for Phase 2 upgrades. It came back to 206 Squadron as 'A' before going to Langar for Phase 3 upgrades in June 1963. In June 1966, WR985 was delivered to Hawker Siddeley Aviation at Woodford for the installation of the additional underwing Viper engines before being delivered to Kinloss with 201 Squadron as 'H'. At the end of its service life, WR985 was transferred to Cosford for use as an instructional airframe with No 2 SoTT, where it was allocated the maintenance serial 8103M upon arrival. WR985 was yet another airframe offered for sale by the MoD in 1988 and was sold to a private purchaser who moved the aircraft to Long Marston. Here it remains today, although it is in appalling condition and in immediate need of some serious TLC if the aircraft is not to fall apart where it sits!

Cyprus

The island of Cyprus boasts no fewer than four Shackletons, although – sadly – the condition of the aircraft varies from poor to derelict!

The first to arrive was MR.3, XF700. It had made its first flight on 9 July 1958 and had been delivered to 120 Squadron as 'A' in September. It continued with its service career until being retired in October 1971. It was despatched to Nicosia on 26 October, to be scrapped for spares recovery and the remains utilised in fire-fighting training. No 103 MU undertook the spares recovery, which was completed by January 1972, and the remains were left to the elements. The airframe and inner wings (minus engines) lie abandoned at the now-disused Nicosia Airport.

The other three Shackletons in Cyprus are all former 8 Squadron aircraft: WL747, WL757 and WR967 have been abandoned at the end of the runway at Paphos International Airport.

WL747 made its first flight as an MR.2 on 5 February 1953 and entered squadron service soon afterwards. In December 1970, the aircraft was flown to 5 MU for a major overhaul ahead of its delivery to Hawker Siddeley Aviation at Bitteswell in February 1971, becoming the very first AEW.2 conversion. It made its maiden flight as an AEW.2 on 2 February 1971 and, after the fitting of its 'new' radar at Lossiemouth, was flown to Kinloss in April 1972 where it joined 8 Squadron and received the name *Florence*.

WL757 made its first flight as an MR.2 on 10 April 1953. It was allocated to the MEAF and joined 37 Squadron as 'D' in August. It remained in service with a number of squadrons before returning to Hawker Siddeley Aviation at Bitteswell in August 1971 for AEW.2 conversion. It made its first flight in the new configuration in June 1972. It was delivered to Lossiemouth for radar installation in August and then transferred to Kinloss to join 8 Squadron, adopting the name *Brian*. It remained with 8 Squadron for the rest of its service life before being retired with almost 14,000 hours on its airframe.

WR967 made its first flight as an MR.2 on 17 May 1954. It was delivered to 23 MU in May 1954 and placed into storage before transferring to the JASS Flight in January 1955. After suffering some damage following

ABOVE Former 8 Squadron Shackleton AEW.2, WL747, is one of two complete aircraft abandoned at Paphos International Airport (the other is WL757). Also abandoned here is the former 8 Squadron *Dodo*. Having been left outside in close proximity to the sea, they have all suffered badly and are now in very poor condition. This image was taken on 4 June 2004.
(Barrie Lewis)

a tailwheel collapse on landing, WR967 was repaired and rejoined the JASS Flight at Ballykelly. It remained in squadron use until flown to Bitteswell in March 1967 for conversion to T.2 configuration. It returned to Ballykelly in September 1968 and operated with the MOTU as 'Q' before suffering damage in May 1969. The aircraft was repaired by 71 MU and it was sent back to Ballykelly, only to suffer further damage which was subsequently repaired.

Next stop for WR967 was Sharjah where the aircraft operated with 210 Squadron. The aircraft arrived back in the UK in November 1971 and was allocated to the RAF Fire School at Catterick, but the allocation was cancelled and the aircraft was delivered to RAF Lossiemouth ahead of the re-formation of 8 Squadron. It was named *Zebedee* and was used for crew training duties until damaged in a flying accident on 7 September 1972, which grounded the aircraft.

In 1974, it was decided to convert the fuselage of WR967 into an AEW training aid. The work was undertaken by Marconi Elliot Avionic Systems Ltd, with work commencing in July that year. The wings and tail unit were removed and the fuselage mounted on to blocks, the 'simulator' being handed over to the RAF in August 1975, when it was appropriately named *Dodo* after the extinct flightless bird. The maintenance serial 8398M was issued, but this actually appeared as 'T83987' on the fuselage. The *Dodo* continued in use with 8 Squadron until the retirement of the Shackleton in July 1991.

When the remaining Shackleton aircraft from 8 Squadron's fleet came up for disposal and were offered for tender, some were delivered to museums and flying organisations around the UK. A Cypriot individual made a successful bid for two aircraft – WL747 and WL757 – and these two aircraft were flown to Paphos with a view to keeping at least one of them in the air. As part of the bid, he also received the *Dodo*, which was shipped to Paphos. Sadly, absolutely nothing came of it and the aircraft remain at Paphos Airport today, in appalling condition, having been left out in the elements around 300m from the sea. Recent investigations of the airframes have showed that little, if anything, is salvageable. What a complete waste!

South Africa

South Africa provided the only export version of the Shackleton when they ordered eight MR.3 aircraft. These were initially in Phase 1 form, later upgraded to Phase 2 and eventually to Phase 3. However, for a number of reasons, the SAAF aircraft were not fitted with the additional underwing Viper jets – one of the main reasons being that the 6,000ft runway requirements of the RAF did not apply in South Africa. With hindsight, this proved to be a beneficial decision for the SAAF as less strain was imposed on their airframes without the extra engines.

The first two SAAF Shackletons (1716 and 1717) were handed over at Woodford

BELOW After being withdrawn from service, SAAF Shackleton MR.3, 1720/M, was placed on display at Ysterplaat AFB but suffered badly from corrosion not apparent in this image from 2002. Sadly, the decision to scrap the aircraft was finally made in March 2013. *(Peter R. March)*

LEFT SAAF
Shackleton MR.3,
1716/J, was restored
to flying condition by
volunteers in 1994 and
accepted an invitation
to tour a number of
airshows in the UK.
While en route to the
UK, both starboard
engines failed and the
crew were left with
no alternative than
a forced landing on
to the desert below,
which the pilots of
'Pelican 16' managed
successfully. None
of the 19 people on
board were injured.
Today, the wreckage
of 'Pelican 16' remains
where she came to
rest on the night of
13 June 1994. *(Alexei
Shevelev)*

on 21 May 1957. By February 1958, eight
aircraft (1716–1723 with code letters 'J' to 'Q')
had been delivered to the SAAF where they
operated from the D.F. Malan Airport near Cape
Town, adjacent to Ysterplaat AFB.

The aircraft had a distinguished service
record until marred by the crash of 1718 on
8 August 1963 when all 13 crew were killed.

Maintaining the Shackletons proved
difficult for the SAAF, particularly during the
embargo imposed by the United Nations
over South Africa's policy of apartheid.
Acquiring components for the Shackleton fleet
became increasingly difficult and the aircraft's
serviceability suffered. The lack of engine spares

and tyres – together with airframe fatigue –
took a gradual toll. By November 1984, the
fatigue lives of all but the two re-sparred aircraft
had expired and the fleet was retired into
storage. However, that was not to be the end of
the story!

Restored to flying condition by volunteers
in 1994, 1716 – callsign 'Pelican 16' – was
invited to take part in a number of airshows
in the UK. The aircraft departed South Africa
for England on 12 July 1994. At the time she
was the only airworthy Shackleton MR.3 in the
world. Flown by a group of active SAAF pilots,
'Pelican 16' was operating over the Sahara
in temperatures exceeding 38°C on the night

LEFT After being
withdrawn from
service, SAAF
Shackleton MR.3,
1721/N, was put on
display at the South
African Air Force
Museum at Swartkop
AFB, where it remains
today. *(Martin Pole)*

ABOVE Technically, SAAF MR.3 1722/P was the last 'airworthy' Shackleton, having been maintained in flying condition by the South African Air Force Historic Flight at Ysterplaat AFB. It has since been grounded for 'safety and preservation reasons', as well as a lack of qualified crew to fly her. *(Ray Barber)*

of 13 July when her No 4 engine began to overheat from a coolant leak and had to be shut down. Moments later, a bolt connecting the two contra-rotating propellers in her No 3 engine failed, causing the assembly to overheat and melt. Left no option but a controlled ditching in the desert below, the pilots of 'Pelican 16' successfully belly-landed their aircraft on flat sands where it slid to a stop. None of the crew was injured in the landing, although all 19 men were miles from any assistance and in the middle of an active war zone. Through the next several days, the crew of 'Pelican 16' made their way to the safety of friendly forces and returned to South Africa. Today, the wreckage of 'Pelican 16' remains where she came to a stop on the night of 13 June 1994.

Other SAAF Shackletons have been preserved and are on display to the public.

No 1721/N is preserved at the South Africa Air Force Museum at Swartkop AFB and the aircraft is displayed outside. Meanwhile, the last 'airworthy' Shackleton MR.3 – 1722/P – is maintained by the South African Air Force Historic Flight at Ysterplaat AFB. However, although it remains technically 'airworthy', it has since been grounded for 'safety and preservation reasons', as well as a lack of qualified crew to fly her. Apparently, there are very few hours left on the airframe's fatigue life; too few to allow a crew to be trained and kept current while still displaying it.

Finally, there is a SAAF example – 1723 – which was acquired by a private individual and placed on display at Vic's Viking Garage, next to the Golden Highway in Meredale, Johannesburg. Having replaced Viking ZS-DGH on display on the garage roof, 1723 was

RIGHT Former SAAF Shackleton MR.3, 1723, was acquired by a private individual and was placed on display at Vic's Viking Garage in Johannesburg. It is seen here in this striking Coca-Cola colour scheme, but subsequent to this image, 1723 has carried a variety of interesting colour schemes for a number of sponsors. *(Martin Pole)*

painted in a striking red Coca-Cola colour scheme. Since then, the aircraft has had at least two other colour schemes applied: a light sand scheme painted as Airtime Lady, and finally a blue, white and gold scheme for Sasol.

United States

As noted above, WL790 departed the UK in 1994 and was flown to the USA where it continued to grace the skies for a further 14 years with the Commemorative Air Force at Midland, Texas, as N790WL. Later, the aircraft underwent a complete respray into its original 8 Squadron colours and was rolled out to the public on 20 May 2013. The aircraft is now a ground exhibit at the Pima Air & Space Museum in Tucson, Arizona, and sadly is not expected to fly again.

WL790 made its first flight as an MR.2 on 23 June 1953 and spent the next 18 years in squadron service. It returned to Bitteswell for conversion to AEW.2 configuration in September 1971 and was sent to Lossiemouth in September 1972 for the installation of the radar. It was delivered to 8 Squadron at Kinloss later in the same month. Here it remained until the type was withdrawn from service and replaced by the Sentry. As of 1989, WL790 had almost 14,000 hours on the airframe – an amazing feat!

The aircraft was sold at auction and moved to Coventry, before finding its way across the Atlantic where it continued to impress American flying show audiences with its amazing roar. Sadly, it too is now quiet.

A full listing of extant Shackleton aircraft can be found in Appendix 4.

ABOVE After leaving the UK for the USA in 1994, WL790 continued to fly for a further 14 years – providing US audiences with the spectacular sight and sound of the amazing 'Growler'! *(Peter R. March)*

LEFT Now, having been completely restored and resprayed, WL790 has been prepared as a ground exhibit at the Pima Air & Space Museum in Tucson, Arizona. She is not expected to fly again. *(Peter R. March)*

Appendix 1

The Shackleton Preservation Trust

The primary aim of the Shackleton Preservation Trust (SPT) is to return Avro Shackleton MR.2/AEW.2, WR963, to flight as an enduring flying memorial to all Shackleton air- and groundcrews from the period 1949 to 1991, as well as being an educational aid to ensure future generations learn about the Shackleton's vital role.

The current chairman is Dave Woods. Dave was initially talked into visiting the Shackleton back in 2009 when his son Rich showed an interest in supporting WR963. Once at Coventry, Dave met the then chairman, John Cubberley, and they hit it off straight away. The initial plan was to visit and help perhaps once a month . . . this soon became twice a month and more recently it involves most weekends for both Dave and Rich.

Due to health issues, John Cubberley planned to stand down and an approach was made to Dave Woods to take on the role of chairman – which he did around 4 years ago.

Dave's view of the SPT is clear: while they are actively involved in trying to get WR963 back into the air, they also exist to provide assistance to any Shackleton in need of help – especially with spare parts as the Trust has a good supply of them. In the past, assistance hasn't stopped at Shackleton owners, as the Battle of Britain Memorial Flight recently approached them for spare parts to keep their Lancaster PA474 in the air; parts which the SPT were happy to assist with.

However, the main objective is to get WR963 back into the air. When a feasibility study was completed soon after the SPT's acquisition of WR963, an amount estimated in the region of £3–4 million was anticipated. However, this assumed that a major re-sparring of the aircraft would be required. Research by the recent trustees indicated that this amount may actually be significantly less – at perhaps £1–2 million. Currently the SPT benefits from four good sponsors, but Dave admits 'One good sponsor investing a decent amount of cash would kick-start the project. The manpower is there, as is a good supply of spare parts including six spare Griffon engines.'

Currently, the SPT owns the design rights to the Avro Shackleton and does not, therefore, need the assistance of BAE Systems. This is just as well, since BAE Systems have already stated in writing their unwillingness to support the project!

A recent arrival with the team and its newest trustee is Druid Petrie, a former 8 Squadron engineer with almost nine years' experience on the Shackleton. His engineering enthusiasm, knowledge and professionalism will help the Trust get WR963 back into the air.

The SPT has a wonderful group of volunteers – all with very differing backgrounds and skills – and, as Dave puts it, 'are all vitally important to the overall project. Most don't have an

BELOW The Shackleton Preservation Trust's WR963 photographed during one of its engine runs at Coventry in March 2015. Chairman Dave Woods is seated in the first pilot's position looking towards the Nos 1 and 2 engines.
(Keith Wilson)

aviation background, but it seems that many do have a background in classic cars and motorcycles and slot in quickly.' If people would like to volunteer to work on the Shackleton, they should contact Dave Woods via the SPT website or Facebook page.

Most maintenance work is carried out to WR963 over the winter, but engine runs are also planned at least once a month during the spring and summer. They may be even more frequent if the aircraft can be made to taxi around the airport. When this occurs, seats on board the aircraft will be offered during the taxi run as part of a day-long visit to WR963. Seat prices will vary but as the Shackleton is an expensive aircraft to taxi around (fuel costs of £600–800 for each taxi run are anticipated) the prices will reflect this. However, what a unique opportunity it would be!

Access to WR963 is permitted to members of the public on Saturdays from 10.00am. The SPT do, however, ask to be informed of your visit in advance – owing to security arrangements at Coventry Airport – and making sure one of the team can be there to grant access on to the airport and on to the aircraft itself. However, with the imminent return of Airbase to Coventry, access to the site should become a little easier. If you are planning to visit, enter the airport via the West Gate, which is located on Stoneleigh Road, Baginton. Then follow the signs for the DC-6 Diner, which will get you to the car park; from there you will be able see WR963.

Not unsurprisingly, donations to the Trust are always welcome, which can be made through the website or Facebook page. Meanwhile, there is also an opportunity to become a 'Friend of WR963'. Membership involves a one-off joining fee of £5, plus an annual membership of just £10.

The SPT is very much an inclusive organisation which welcomes all volunteers. As Dave Woods makes absolutely clear, 'While the aircraft is the property of the Trust and we are bound by the Charity Commission rules, the aircraft's continued survival is due to those who put their heart and soul into it.'

It is not now a matter of 'if' WR963 flies again, but 'when'!

Website: **www.avroshackleton.co.uk**
Facebook page: Avro Shackleton – Non-profit Organisation
Email: **shackletonwr963@googlemail.com**
Twitter: @WR963

Appendix 2

Avro Shackleton – technical specifications

	MR.1/MR.1A	MR.2/MR.2C	MR.3	T.4	AEW.2
Dimensions:					
Span	120ft 0in	120ft 0in	119ft 10in	120ft 0in	120ft 0in
Length	77ft 6in	87ft 3in	92ft 6in	77ft 6in	87ft 3in
Height	17ft 6in	17ft 6in	23ft 4in	17ft 6in	17ft 6in
Wing area	1,421sq ft	1,421sq ft	1,458sq ft	1,421sq ft	1,421sq ft
Weights:					
Empty weight	49,600lb	51,400lb	64,300lb	49,600lb	58,800lb
All-up weight	86,000lb	86,000lb	100,000lb	86,000lb	96,100lb
Usable load	36,400lb	34,600lb	35,700lb	36,400lb	37,300lb
Performance:					
Maximum speed @ 12,000ft	294mph	286/299mph	297mph	294mph	262mph
Maximum cruise @ 10,000ft	245mph	249/249mph	253mph	245mph	
Initial rate of climb	1,005ft/min	900/920ft/min	850ft/min	1,005ft/min	900ft/min
Service ceiling	20,700ft	18,800/20,200ft	18,600ft	20,700ft	
Range (at 5,000ft at 190mph)	2,160nm	2,780/2,900nm	3,660nm	2,160nm	
Max still-air range	3,080nm	2,780/2,900nm	3,660nm	3,080nm	2,700nm
Endurance	14.8 hours	13/14.6 hours	16 hours	14.8 hours	13 hours
Landing speed	114mph	116/114mph	118mph	114mph	116mph
Stalling speed	88mph	92/88mph	96mph	88mph	92mph
Powerplants:	Four × 2,435hp Rolls-Royce Griffon 57/57A engines	Four × 2,435hp Rolls-Royce Griffon 57A engines	Four × 2,435hp Rolls-Royce Griffon 57A engines	Four × 2,435hp Rolls-Royce Griffon 57/57A engines	Four × 2,435hp Rolls-Royce Griffon 58 engines
			Two × Rolls-Royce Viper 203 powerplants were added after the Phase 3 modifications		
Propellers:	Four × de Havilland 13ft-diameter contra-rotating, constant-speed, fully feathering units	Four × de Havilland 13ft-diameter contra-rotating, constant-speed, fully feathering units	Four × de Havilland 13ft-diameter contra-rotating, constant-speed, fully feathering units	Four × de Havilland 13ft-diameter contra-rotating, constant-speed, fully feathering units	Four × de Havilland 13ft-diameter contra-rotating, constant-speed, fully feathering units
Fuel capacity (gallons):	3,292	3,292/3,350	4,316	3,292	3,350
Total number produced:	32/48	69	34 (+8 for SAAF)	17 conversions	12 conversions

Appendix 3

Shackleton production list

Serial	Mark	First flight	Details	Fate
VW126	MR.1	09.03.1949	First prototype. Built to Specification R5/46. Converted to MR.2	No 2 Radio School as 7626M 02.1960. Broken up at Yatebury, 10.1965
VW131	MR.1	02.09.1949	Second prototype. Napier Nomad test-bed	Tested to destruction. SOC 10.05.1962
VW135	MR.1	29.03.1950	Third prototype. Retained by manufacturer for extended trials work	Broken for spares. SOC 13.04.1954
VP254	MR.1	28.03.1950	Built to Specification 42/46. Manufacturer's trials, Langar – IFF10 and SARAH trials	Crashed into South China Sea 09.12.1958
VP255	MR.1	30.06.1950	Built to Specification 42/46	Sold for scrap 22.08.1963
VP256	MR.1	18.09.1950	Built to Specification 42/46	SOC 14.04.1955 at 23MU. Sold for scrap 02.1963
VP257	MR.1	28.08.1950	Built to Specification 42/46	NEA, sold for scrap 28.02.1963
VP258	MR.1	13.10.1950	Built to Specification 42/46. Converted to T.4 at Woodford 05.1955	To Stansted Fire School 17.07.1968. Burnt
VP259	MR.1	24.10.1950	Built to Specification 42/46. Converted to T.4 28.04.1956	Crashed Haldon Hill, Elgin 10.01.1958
VP260	MR.1	22.01.1951	Built to Specification 42/46	NEA, broken up 1962. Sold for scrap 29.09.1963
VP261	MR.1	13.02.1951	Built to Specification 42/46	Crashed in sea near Berwick-upon-Tweed 25.06.1952
VP262	MR.1	20.02.1951	Built to Specification 42/46	NEA, broken up 1962. Sold for scrap 22.08.1963
VP263	MR.1	17.03.1951	Built to Specification 42/46. Grapple mods	NEA, sold for scrap 22.08.1963
VP264	MR.1	08.03.1951	Built to Specification 42/46	Sold for scrap 22.11.1962
VP265	MR.1	29.03.1951	Built to Specification 42/46	NEA, sold for scrap 23.10.1963
VP266	MR.1	04.04.1951	Built to Specification 42/46	NEA, sold for scrap 28.02.1963
VP267	MR.1	13.04.1951	Built to Specification 42/46	Reduced to components 01.12.1962
VP268	MR.1	20.04.1951	Built to Specification 42/46	NEA, sold for scrap 23.10.1963
VP281	MR.1	24.04.1951	Built to Specification 42/46	NEA, sold for scrap 28.02.1963
VP282	MR.1	01.05.1951	Built to Specification 42/46. Orange Harvest trials	NEA, sold for scrap 31.05.1962
VP283	MR.1	11.05.1951	Built to Specification 42/46	Struck off charge after landing accident, Gibraltar 12.8.1951
VP284	MR.1	23.05.1951	Built to Specification 42/46	NEA, sold for scrap 28.02.1963
VP285	MR.1	26.05.1951	Built to Specification 42/46. 'Blue Silk' trials	NEA, broken up, sold for scrap 28.02.1963
VP286	MR.1	31.05.1951	Built to Specification 42/46	Crashed into sea near Cromarty 8.10.1952
VP287	MR.1	07.06.1951	Built to Specification 42/46	NEA, sold for scrap 23.10.1963
VP288	MR.1	15.06.1951	Built to Specification 42/46	NEA, sold for scrap 05.08.1964
VP289	MR.1	25.06.1951	Built to Specification 42/46. Grapple mods	To 7730M. Scrapped 04.1966
VP290	MR.1	25.06.1951	Built to Specification 42/46.	NEA, sold for scrap 22.11.1962
VP291	MR.1	29.06.1951	Built to Specification 42/46	NEA, sold for scrap 05.08.1964
VP292	MR.1	12.07.1951	Built to Specification 42/46	NEA, Category 5 components, SOC 28.04.1961
VP293	MR.1	18.07.1951	Built to Specification 42/46. Converted to T.4	Sold to Strathallan Museum 13.05.1976. Broken up on site. Cockpit section preserved at Coventry
VP294	MR.1	18.07.1951	Built to Specification 42/46	Damaged beyond repair on landing at Gan, 15.05.1962; Category 5 components, SOC 18.05.1962
WB818	MR.1A	01.08.1951		Suffered taxiing accident at Gan, 20.5.1961. Later flown to Seletar and placed in storage. Category 5 components and SOC 28.04.1962
WB819	MR.1A	02.08.1951	Converted to T.4 13.06.1957	SOC, Stansted Fire School 14.06.1968. Burnt

Serial	Mark	First flight	Details	Fate
WB820	MR.1A	14.08.1951	Converted to T.4 08.09.1960	Category 5 components, SOC 01.06.1967. St Mawgan dump
WB821	MR.1A	17.08.1951		NEA, sold for scrap 31.05.1962
WB822	MR.1A	17.08.1951	Converted to T.4 08.11.1960	Category 5 components, SOC, fire practice 08.1968. Burnt
WB823	MR.1A	25.08.1951		NEA, sold for scrap 29.05.1963
WB824	MR.1A	30.08.1951		NEA, sold for scrap 03.01.1962
WB825	MR.1A	31.08.1951		Category 5 components, SOC 08.08.1961. Sold for scrap
WB826	MR.1A	03.09.1951	Grapple mods 01.1957	NEA, sold for scrap 20.02.1968
WB827	MR.1A	12.09.1951		Sold for scrap 05.08.1954
WB828	MR.1A	14.09.1951		NEA, sold for scrap 22.11.1962
WB829	MR.1A	21.09.1951		NEA, Category 5 components, SOC 28.04.1962
WB830	MR.1A	26.09.1951		NEA, sold for scrap 31.05.1962
WB831	MR.1A	27.09.1951	Converted to T.4 05.02.1956	Sank back on to runway following premature landing gear retraction at the MOTU, St Mawgan, 17.05.1967. Category 5 components, SOC 07.06.1967. St Mawgan fire dump. Burnt
WB832	MR.1A	03.10.1951	Converted to T.4 23.08.1956	To 7885M 08.07.1965. Category 5 components. Scrapped
WB833	MR.1A	17.06.1952	Converted to prototype Shackleton MR.2	Crashed on the Mull of Kintyre, 19.04.1968. SOC
WB834	MR.1A	11.10.1951		Category 5 components. SOC 08.08.1961
WB835	MR.1A	15.10.1951	Manufacturer's trial with Mk 3 lifeboat	NEA, sold for scrap 23.10.1963
WB836	MR.1A	18.10.1951	Grapple mods	NEA, sold for scrap 05.08.1964
WB837	MR.1A	24.10.1951	Converted to T.4 28.03.1956	NEA, sold for scrap 03.02.1969
WB844	MR.1A	31.10.1951	Converted to T.4 10.07.1956	To 8028M 30.07.1968. Scrapped at Cosford
WB845	MR.1A	08.11.1951	Converted to T.4 18.10.1960	NEA, sold for scrap 12.03.1969
WB846	MR.1A	14.11.1951		Withdrawn 26.03.1958. To MOTU as 7561M. Scrapped Kinloss
WB847	MR.1A	20.11.1951	Converted to T.4 22.08.1956	Withdrawn 06.1968 to 8020M. Allocated to 'gate guardian' at Kinloss but cancelled. Fire dump 12.03.1969. Scrapped
WB848	MR.1A	25.11.1951		NEA, sold for scrap 23.10.1963
WB849	MR.1A	28.11.1951	Converted to T.4 04.08.1960	To Newton as 8026M 30.07.1968. Category 5 components. Scrapped
WB850	MR.1A	08.12.1951		NEA, sold for scrap 29.05.1963
WB851	MR.1A	12.12.1951		NEA, sold for scrap 28.02.1963
WB852	MR.1A	14.12.1951		NEA, sold for scrap 29.05.1963
WB853	MR.1A	19.12.1951		NEA, sold for scrap 29.05.1963
WB854	MR.1A	29.12.1951	Conversion to T.4 cancelled 02.1957	Withdrawn, Category 5 components. SOC 27.11.1962. Scrapped Seletar
WB855	MR.1A	02.01.1952		NEA, sold for scrap 28.02.1963
WB856	MR.1A	09.01.1952	Grapple mods	NEA, sold for scrap 19.12.1960
WB857	MR.1A	16.01.1952	Grapple mods 11.1955	27 MU storage 27.11.1959. Sold for scrap 31.05.1962
WB858	MR.1A	29.01.1952	Converted to T.4 31.08.1960	Withdrawn 27 MU 12.07.1968. NEA, sold for scrap 03.02.1969
WB859	MR.1A	30.01.1952		23 MU storage 18.02.1959. Sold for scrap 25.09.1963
WB860	MR.1A	07.02.1952	Grapple mods 10.1956	23 MU storage 18.03.1960. NEA, sold for scrap 28.02.1963
WB861	MR.1A	20.02.1952	Grapple mods 25.06.1956	Crash landing 05.09.1957. Category 5 components. SOC 06.09.1957
WG507	MR.1A	26.02.1952		NEA, scrapped 31.05.1962
WG508	MR.1A	08.03.1952		NEA, sold for scrap 29.05.1963
WG509	MR.1A	13.03.1952	Grapple mods 11.1956	Sold for scrap 25.09.1963

Serial	Mark	First flight	Details	Fate
WG510	MR.1A	30.03.1952		NEA, sold for scrap 28.02.1963
WG511	MR.1A	07.04.1952	Converted to T.4 25.05.1956	Category 5 components 26.07.1966. SOC 03.08.1966. No 71 MU conversion to front fuselage 1968–71. RAF Museum 08.1974. Sold to Cornwall Aero Park. Cockpit section on display with the Flambards Experience at Helston
WG525	MR.1A	18.04.1952		NEA, sold for scrap 05.08.1964
WG526	MR.1A	06.05.1952		Category 5 components. SOC. Scrapped 11.07.1961
WG527	MR.1A	15.05.1952	Converted to T.4 18.07.1956	NEA, sold for scrap 12.03.1969
WG528	MR.1A	18.05.1952		NEA, sold for scrap 29.05.1963
WG529	MR.1A	24.06.1952		NEA, broken up 1962. Sold for scrap 25.11.1963
WG530	MR.2	15.08.1952	A&AEE tropical trials	NEA, sold for scrap 03.09.1968
WG531	MR.2	21.08.1952	SBAC display 01.09.1952 to 07.09.1952	Believed to have collided with WL743 on 11.01.1955. Missing
WG532	MR.2	12.09.1952	Glow Worm rocket flare trials	NEA, sold for scrap 03.09.1968
WG533	MR.2	18.09.1952	Converted to T.2 20.02.1967	NEA, for scrap sale 10.12.1973. Cancelled, St Athan dump
WG553	MR.2	02.10.1952		NEA, sold for scrap 24.06.1968
WG554	MR.2	10.10.1952	Converted to T.2 16.01.1967	NEA, Category 5 components 16.11.1972. St Athan dump
WG555	MR.2	21.10.1952		SOC, RAF Fire Fighting School, Catterick, 09.05.1972. Burnt
WG556	MR.2	28.10.1952	RAE and ASWDU Jezebel trials	Category 5 damage 1980. BDRF Lossiemouth, then Fire Section as 8651M 20.02.1981. Scrapped 07.1982
WG557	MR.2	05.11.1952	RAE Armament Flight and ETPS	SOC 28.10.1964. Farnborough dump
WG558	MR.2	11.11.1952	Conversion to T.2 30.11.1966	NEA, Category 5 components 25.01.1974. Scrapped
WL737	MR.2	17.11.1952	MoA glidepath trials. A&AEE high all-up weight trials	NEA, Category 5 components 31.08.1973. St Athan dump. Scrapped
WL738	MR.2	25.11.1952	A&AEE radio altimeter trials. 8 Squadron pilot training 08.03.1974	Category 5 components, SOC 14.10.1977. Cancelled, to 8567M and 'gate guard' at RAF Lossiemouth 05.04.1978. Scrapped 1991
WL739	MR.2	09.12.1952	Conversion to T.2	RAF Fire School, Catterick, 01.11.1971. Cancelled, to Manston Fire Fighting School 1971. Burnt
WL740	MR.2	18.12.1952	A&AEE gun trials	Conversion to T.2 not completed. Category 5 components, SOC 28.02.1968
WL741	MR.2	18.12.1952	Converted to AEW.2 04.1972 and named *P.C. Knapweed*	Withdrawn 09.01.1981. FSCTE Manston as 8692M 09.01.1982. Burnt 06.1982
WL742	MR.2	23.12.1952		NEA, sold for scrap 26.06.1968
WL743	MR.2	20.01.1953		Believed to have collided with WG531 on 11.01.1955. Missing
WL744	MR.2	15.01.1953		Category 5 components 20.10.1966. SOC, RAF Ballykelly dump 01.11.1966. Scrapped
WL745	MR.2	22.01.1953	Woodford performance trials with AN/APS 20 radar 03.1970. Converted to AEW.2 04.1972. Named *Sage*	Withdrawn from service 06.1981. RAF Fire Fighting School, Catterick, as 8698M 13.07.1981. Burnt
WL746	MR.2	28.01.1953		Crashed into sea off Argyll, 11.12.1953. Salvaged. SOC Category 5 scrap 12.12.1953
WL747	MR.2	05.02.1953	Converted to AEW.2 02.02.1971. Named *Florence*	Withdrawn from service 07.1991. Sold to Savvas Constantinides 03.07.1991. Delivered to Paphos Airport 19.07.1991. Derelict at Paphos
WL748	MR.2	06.02.1953	RRE Pershore infrared scanner trials	RAF Fire Fighting School, Catterick, 08.05.1972. Burnt
WL749	MR.2	18.02.1953		Suffered Category 5 damage 14.05.1953. Scrap, components 14.05.1953
WL750	MR.2	23.02.1953	Conversion to T.2 11.01.1967 to 27.03.1968	RAF Fire Fighting School 15.10.1971. Cancelled, FSCTE Manston 11.1971

Serial	Mark	First flight	Details	Fate
WL751	MR.2	05.03.1953	HAS stall warning system trials	Sold Shackleton Aviation, Baginton, 04.05.1972. Sold forscrap 01.1975
WL752	MR.2	07.03.1953		NEA, 13.09.1967. Sold for scrap 07.10.1968
WL753	MR.2	16.03.1953		NEA, broken up 25.05.1967. Sold for scrap 12.03.1969
WL754	MR.2	18.03.1953	Bitteswell ADD stall warning trials. Converted to AEW.2 22.03.1972. Named *Paul*	Withdrawn from service 01.1981. Fire practice, crash rescue, RAF Valley as 8665M 22.01.1981. Display 'Save the Shackleton' campaign
WL755	MR.2	30.03.1953		RAF Fire School, Catterick, 09.05.1972. Burnt 12.77
WL756	MR.2	01.04.1953	Converted to AEW.2 02.04.1972. Named *Mr Rusty*	Withdrawn from service 01.07.1991. Crash rescue RAF St Mawgan 07.91. Burnt by 1998. Cockpit section on display at Caernarfon Airport Airworld Museum
WL757	MR.2		Converted to AEW.2 26.08.1971. Named *Brian*	Withdrawn from service 07.1991. Sold to Savvas Constantinides 03.07.1991. Delivered Paphos Airport 15.07.1991. Derelict at Paphos
WL758	MR.2	17.04.1953		RAF Fire School, Catterick, 19.05.1972. Burnt by 1975
WL759	MR.2	24.04.1953	Mk.44 torpedo trials 22.05.1963	SOC, Category 5 components 11.11.1968. Broken up 1969
WL785	MR.2	10.05.1953	A&AEE sonobuoy trials 12.04.1965 to 27.05.1965	FSCTE Manston 19.05.1971. SOC, Category 5 components 30.06.1971
WL786	MR.2	12.05.1953		Crashed into Indian Ocean off Indonesia 05.11.1967. SOC 06.11.1967. Missing
WL787	MR.2	18.05.1953	Conversion to T.2 14.12.1966. Training equipment removed 02.06.1970	RAF Fire School, Catterick, cancelled 01.11.1971. No8 Squadron crew training 01.01.1972. Named *Mr McHenry*. Category 3 damage and repaired on site. Returned to 8 Squadron and renamed *Dylan* 15.05.1973. Withdrawn from service, firefighting practice 03.01.1974. Broken up and destroyed 03.1974
WL788	MR.2	04.06.1953		NEA, SOC, Category 5 components 12.09.1967. Sold for scrap 28.03.1969
WL789	MR.2	10.06.1953	MAD tail boom installation 08.1953. ASWDU MAD trial 09.1953	NEA, Category 5 components, broken up 12.1968. Sold for scrap 28.03.1968
WL790	MR.2	23.06.1953	Converted to AEW.2 30.09.1971. Named *Mr McHenry* initially but later *Zebedee*	Withdrawn from service 01.07.1991. To Air Atlantique for storage at Baginton 10.07.1991. Sold to Polar Aviation Museum, Minnesota, and delivered as N790WL 07.09.1994. Now preserved with the Pima Air & Space Museum at Tucson, Arizona, as N790WL
WL791	MR.2	16.06.1952		NEA, Category 5 scrap 23.05.1967. Broken up 12.68. Sold for scrap 28.03.1969
WL792	MR.2	01.07.1953		Crash-landing during air display at RAF Gibraltar 14.09.1957. Category 5 components 11.11.1957
WL793	MR.2	15.07.1953	Converted to AEW.2 05.06.1972. Named *Ermintrude*	Withdrawn from service 1981. BDRF RAF Lossiemouth as 8675M, but dumped 07.1981. Scrapped 07.1982
WL794	MR.2	07.08.1953		Crashed into Mediterranean Sea off Gozo 12.02.1954. Category 5. Missing
WL795	MR.2	17.08.1953	Converted to AEW.2 04.02.1972. Named *Rosalie*	Withdrawn from service 1981. Fire practice and crash rescue St Mawgan as 8753M 24.11.1981. Cancelled, stored St Mawgan, then refurbished for display at St Mawgan 04.1989. Currently being disposed of
WL796	MR.2	23.08.1953		NEA, 01.11.1967. Sold for scrap 07.10.1968.
WL797	MR.2	15.09.1953		NEA, 12.09.1967, Category 5 components. sold for scrap 07.10.1968
WL798	MR.2	17.09.1953		To 2 SoTT Cosford as 8114M 04.12.1970. Lossiemouth spares for 8 Squadron
WL799	MR.2	18.09.1953		Destroyed in hangar fire at Langar on 22.12.1955
WL800	MR.2	01.10.1953		RAF Fire Fighting School, Catterick, and SOC 19.05.1972. Burnt

Serial	Mark	First flight	Details	Fate
WL801	MR.2	10.10.1953	No 8 Squadron crew training 15.08.1974	Withdrawn 06.1979. Aerospace Museum, Cosford. Scrapped 1991
WR951	MR.2	20.10.1953	Converted to T.2 05.04.1967 but cancelled	SOC, Category 5 components 28.02.1968
WR952	MR.2	27.10.1953		NEA, 10.01.1972. Category 5 components 26.09.1973. Broken up
WR953	MR.2	06.11.1953	Langar, permanent oxygen system trials 21.10.1961	NEA, 10.11.1967, fire fighting training, RAF Kinloss, 20.11.1967. SOC, Category 5 components, scrapped 1968
WR954	MR.2	19.11.1953		NEA, 21.09.1967. SOC, Category 5 components 09.05.1973. Scrapped
WR955	MR.2	27.11.1953	A&AEE landing performance trials re AEW.2 30.03.1971 to 04.05.1971	Firefighting training, RAF Brize Norton, 07.06.1971. SOC, Category 5 scrap
WR956	MR.2	10.12.1953		Skidded off runway at Ballykelly on 01.04.1968. SOC, Category 5 scrap, 01.04.1968. Ballykelly dump
WR957	MR.2	21.12.1953		NEA, Category 5 components, 23.05.1967. Sold as scrap 26.06.1968
WR958	MR.2	08.01.1954		NEA, Category 5 components 23.05.1967. Sold as scrap 03.09.1968
WR959	MR.2	20.01.1954		SOC, Category 5 components 03.09.1968. Broken up Changi
WR960	MR.2	05.02.1954	Converted to AEW.2 27.05.1971. Named *Dougal*	Withdrawn from service 11.82. To Cosford as 8772M. Preserved at the Museum of Science and Industry, Manchester, 27.01.1983
WR961	MR.2	12.02.1954		NEA, 23.09.1977. Category 5 components. Sold for scrap 01.02.1978
WR962	MR.2	02.03.1954	A&AEE armamant and Lindholme Gear release trials 25.02.1959	NEA, 13.04.1968. Category 5 components. Sold for scrap 28.03.1969
WR963	MR.2	11.03.1954	Converted to AEW.2 02.06.1972. Named *Parsley* initially and later *Ermintrude*	Currently preserved with the Shackleton Preservation Trust at Coventry, with the AEW.2 radar removed and replaced with its original MR.2 radar configuration.
WR964	MR.2	18.03.1954	A&AEE tropical cooling trials 14.04.1954. Converted to T.2 10.02.1967	NEA, 27.11.1970. Category 5 components. Sold as scrap 15.11.71
WR965	MR.2	07.04.1954	Converted to AEW.2 28.04.1972. Named *Dill* initially and later *Rosalie*	Crashed into Harris on 30.04.1990. SOC 30.04.1990.
WR966	MR.2	28.04.1954	Converted to T.2 26.01.1967	NEA, Category 5 components 22.06.1973. Scrapped
WR967	MR.2	17.05.1954	To 8 Squadron 01.01.1972 and named *Zebedee*	Allocated to RAF Fire Fighting School but posting cancelled. To 8 Squadron 01.01.1972 but suffered Category 3 damage on 07.09.1972. Wings removed and fuselage converted to AEW training simulator with 8 Squadron as 8398M 14.08.1975. Named *Dodo*. Scrapped in 1991 and sold to Savvas Constantinides. Delivered to Paphos Airport. Derelict at Paphos
WR968	MR.2	17.06.1954	A&AEE IFF Mk 10 clearance trials	Crash-landed at Ballykelly and destroyed by fire on 20.10.1961. SOC Category 5 20.10.1961.
WR969	MR.2	10.05.1954	SBAC display aircraft 06.09.1954 to 12.09.1954. Converted to T.2 14.02.1967	NEA, 17.09.1971. SOC Category 5 22.09.1974. Scrapped
WR970	MR.3	02.09.1955	SBAC display aircraft 05.09.1955 to 11.09.1955. A&AEE handling trials 07.09.1956	Crashed at Foolow, Derbyshire, 07.12.1956. SOC 28.11.1957. Scrap
WR971	MR.3	28.05.1956	A&AEE armament trials	To No 2 SoTT, Cosford, as 8119M. Fuselage only sold to Wellesley Aviation, Narborough, 1988. Fuselage on display with Fenland and West Norfolk Aviation Museum near Wisbech
WR972	MR.3	06.11.1956	A&AEE radar, radio, navigational and photographic equipment trials 24.05.1957. Purchased by MoA for RAE 13.03.1959. A7AEE Orange Harvest ECM trials	SOC, Category 5 scrap 31.01.1973. Firefighting and rescue training, Farnborough. Burnt

Serial	Mark	First flight	Details	Fate
WR973	MR.3	18.01.1957	A&AEE Mk 10 autopilot trials	Firefighting practice, RAF Thorney Island, 14.06.1971. SOC, Category 5 scrap 06.1971
WR974	MR.3	01.05.1957	A&AEE tropical and winter trials 5.07.1957	To No 2 SoTT, Cosford, as 8117M 11.12.1970. Sold to Peter Vallance Collection, Charlwood, 1988. Preserved at the Gatwick Aviation Museum at Charlwood but recently dismantled and moved to Bruntingthorpe
WR975	MR.3	26.06.1957	RAF Handling Squadron 07.1957. Viper installation 01.1967	SOC, Category 5 scrap 01.10.1971
WR976	MR.3	19.07.1957	Viper installation 10.1966	Crashed into sea off Land's End 19.11.1967. SOC Category 5 missing 19.11.1967
WR977	MR.3	31.08.1957	SBAC static display 02.09.1957 to 09.09.1957. Viper installation 04.1966	Allocation firefighting practice, RAF Thorney Island, cancelled. To Finningley Museum as 8186M 08.11.1971. Transferred to the Newark Air Museum 01.05.1977. Preserved
WR978	MR.3	09.1957	Viper installation 11.1966	RAF Fire Fighting School, Catterick, 29.11.1970. SOC Category 5 scrap
WR979	MR.3	01.11.1957	Viper installation 12.1968	SOC, Category 5 scrap 01.10.1971. Broken up St Athan
WR980	MR.3	13.11.1957	Viper installation 03.1966	RAF Fire Fighting School, Catterick, 26.11.1970. SOC, Category 5 scrap
WR981	MR.3	12.1957	Viper installation 09.1966	To RAF Topcliffe as 8120M 17.12.1970. Category 5 GI (ground instructional airframe)
WR982	MR.3	02.1958	Viper installation 07.1966. A&AEE Viper, water/methanol and radar altimeter clearance trials	Withdrawn from service 09.1970. To No 2 SoTT, Cosford, as 8106M 06.10.1970. Sold to N. Martin, Lutterworth. Transferred to the Gatwick Aviation Museum at Charlwood. Preserved
WR983	MR.3	03.1958	Viper installation 06.1967	Category 5 scrap 1970. Broken up
WR984	MR.3	06.03.1958	Viper installation 12.1966	To RAF Topcliffe as 8115M 09.11.1970. Category 5 GI (ground instructional airframe) 07.1971
WR985	MR.3	04.1958	A&AEE maritime tactical position indicator trials 22.04.1959. Viper installation 10.1966	To No 2 SoTT, Cosford, as 8103M 25.09.1970. Sold to Jet Aviation Preservation Group at Long Marston. Currently in poor condition at Long Marston aerodrome
WR986	MR.3	04.1958	Phase 3 mods by 09.1966	Category 5 scrap, due to rat infestation 01.09.1971. Broken up at 132 MU
WR987	MR.3	05.1958	Viper installation 12.1966	Fire practice, RAF Honington, 17.06.1972. SOC, Category 5 scrap
WR988	MR.3	05.1958	Viper installation 06.1966	Firefighting practice, Macrihanish, 25.07.1972. SOC, Category 5 scrap
WR989	MR.3	06.1958	Viper installation 04.1967. A&AEE Viper high-humidity and low-temperature flight trials	Firefighting practice, RAF Leeming, 14.07.1972. SOC, Category 5 scrap
WR990	MR.3	07.1958	Viper installation 09.1966	To RAF Newton as 8107M 17.10.1970. SOC, Category 5 GI (ground instructional airframe)
XF700	MR.3	09.07.1958	Viper installation 04.1967	Firefighting practice at Nicosia 26.10.1971. SOC, Category 5 scrap. Spares to 103 MU, then scrapped. Currently, derelict at Nicosia, Cyprus
XF701	MR.3	08.1958	SBAC static display 01.09.1958 to 07.09.1958. Viper installation 04.1966	FSCTE, Manston, 13.08.1971. SOC, Category 5 scrap
XF702	MR.3	09.1958	Viper installation 03.1967	Crashed Creag Bhan, Inverness, on 21.12.1967. SOC 21.12.1967
XF703	MR.3	09.1958	Viper installation 09.1967	To Henlow for RAF Museum 23.09.1971. However, it was scrapped at Henlow in 1975 before it could join the museum.
XF704	MR.3	10.1958	Phase 3 mods by 05.1965	Crashed into Moray Firth 8.12.1965. SOC, Category 5, missing
XF705	MR.3	10.196	Phase 3 mods by 04.1966	Withdrawn from service 20.08.1971. To FSCTE Manston 08.1971. SOC, Category 5, scrap

Serial	Mark	First flight	Details	Fate
XF706	MR.3	12.1958	Viper installation 01.1967	Withdrawn from service 02.1970. Instructional airframe 8089M allocated but not issued. Firefighting practice, St Mawgan 03.1970. Burnt
XF707	MR.3	01.1959	A&AEE cooling of ASV-21 radar assessment 02.1959. Viper installation 01.1968	Withdrawn from service 28.04.1971. Firefighting practice, RAF Benson, 28.04.1971. SOC, Category 5, scrap
XF708	MR.3	01.1959	Phase 3 mods by 06.02.1967	Withdrawn from use and allocated to the Imperial War Museum at Duxford 23.08.1972. Preserved, currently undergoing restoration
XF709	MR.3	03.1959	Viper installation 07.1966. HAS new tailplane and de-icing system trials 06.05.1969	NEA, 30.07.1970. SOC, Category 5, scrap
XF710	MR.3	03.1959	Phase 2 mods by 11.1963	Crash-landed on Culloden Moor, near Inverness, on 10.01.1964. SOC, Category 5, scrap
XF711	MR.3	04.1959	SBAC display, take-off for 22-hour patrol, returning the next day 09.1960. Viper installation 09.1968	Withdrawn from service 06.1971. Firefighting practice, Abingdon, 07.06.1971. SOC, Category 5, scrap
XF730	MR.3	05.1959	Viper installation by 06.1966	Withdrawn from service 24.06.1971. Firefighting practice, Kinloss, 06.1971. SOC, Category 5, scrap
1716	MR.3	29.03.1957	Accepted by 35 Squadron on 16.05.1957	Crash-landed in the Sahara after double engine failure on 13.07.1994 while en route to the UK
1717	MR.3	06.05.1957	Accepted by 35 Squadron on 16.05.1957	Was on display at Transport Museum in Stanger but broken up in 2009.
1718	MR.3	13.05.1957	Accepted by 35 Squadron on 16.05.1957	Crashed into Wemmershoek Mountains 08.08.1963
1719	MR.3	06.09.1957	Accepted by SAAF in 01.1958	Grounded by fatigue hours on 24.04.1978. On display at Cape Town Waterfront but since scrapped
1720	MR.3	26.09.1957	Accepted by SAAF in 01.1958	Grounded by fatigue hours on 10.03.1981 and was placed on display at Ysterplaat AFB until broken up in 2009
1721	MR.3	12.12.1957	Accepted by 35 Squadron on 30.01.1958	On static display at South African Air Force Museum, Swartkop
1722	MR.3	07.02.1958	Accepted by 35 Squadron on 30.01.1958	On display at South African Air Force Historic Flight, Ysterplaat AFB
1723	MR.3	10.02.1958	Accepted by SAAF in 04.02.1958	Grounded by fatigue hours on 22.11.1977. Later placed on display in red colour scheme at Vic's Viking Garage, Meredale, Johannesburg. Had since been repainted into a number of sponsors' colour schemes

Abbreviations:

BDRF   Battle Damage Repair Flight
FSCTE   Fire Service Central Training Establishment
MAD   Magnetic Anomaly Detector
NEA   Non-effective aircraft (at Maintenance Unit)
SOC   Struck off charge

Appendix 4

Shackletons extant

Model	Serial	Code serial	Maintenance markings	Squadron	Location	
United Kingdom						
MR.1/T.4	VP293	X		MOTU	Shackleton Preservation Trust, Coventry Airport, CV8 3AZ	
MR.1A/T.4	WG511				The Flambards Experience, Helston, TR13 0QA	
AEW.2/MR.2	WL795	T	8753M		Inter-Service Survival School, Newquay, TR8 4HP	
AEW.2	WR960		8772M	8 Squadron	Museum of Science and Industry, Manchester, M3 4FP	
AEW.2/MR.2	WR963	B-M		224 Squadron	Shackleton Preservation Trust, Coventry Airport, CV8 3AZ	
MR.3/3	WR971	Q	8119M		Fenland & West Norfolk Aviation Museum, Wisbech, PE14 7DA	
MR.3/3	WR974	K	8117M		Bruntingthorpe LE17 5QS	
MR.3/3	WR977	B	8186M	42 Squadron	Newark Air Museum, Winthorpe, NG24 2NY	
MR.3/3	WR982	J	8106M		Gatwick Aviation Museum, Charlwood, RH6 0BT	
MR.3/3	WR985	H	8103M		Long Marston aerodrome, Stratford-upon-Avon, CV37 8LL	
MR.3/3	XF708	C		203 Squadron	Imperial War Museum, Duxford, CB2 4QR	
Cyprus						
AEW.2	WL747			8 Squadron	Paphos International Airport, Paphos 8061, Nicosia	
AEW.2	WL757			8 Squadron	Paphos International Airport, Paphos 8061, Nicosia	
MR.2/T.2	WR967	'T83987'	8398M	8 Squadron	Paphos International Airport, Paphos 8061, Nicosia	
MR.3	XF700				Nicosia Airport (closed), Lakatamia, Nicosia	
Holland						
AEW.2	WL756		9101M	8 Squadron	Private collector in Holland	
South Africa						
MR.3	1716	J		35 Squadron	Force-landed in Sahara in 1994 and abandoned	
MR.3	1721	N		35 Squadron	South African Air Force Museum, Swartkop AFB, Centurion	
MR.3	1722	P		35 Squadron	South African Air Force Historic Flight, Ysterplaat AFB, Cape Town	
MR.3	1723				Vic's Viking Garage, next to the Golden Highway in Meredale, Johannesburg	
USA						
AEW.2	WL790	N790WL		8 Squadron	Pima Air & Space Museum, Tucson, Arizona	

Condition	Remarks
Cockpit section only	Ex-Duxford, Coltishall, Winthorpe, Coventry, Woodford, Coventry, East Kirkby, Strathallan, RAE, MOTU, A&AEE, [MR.1] 206, 42, 224, 236 OCU. SOC 23.05.75
Cockpit section only	Ex-Colerne, St Mawgan, MOTU, Kinloss Wing, MOTU. SOC 03.08.66
Complete – although recently offered for sale by tender	Ex-8753M, 8, 205, 38, 204, 210, 269, 204 Squadrons
Complete	Ex-Cosford, 8772M, 8, 205, A&AEE, 210, 42, 228 Squadrons
Complete – taxies and may be returned to airworthy condition	Ex-N953WR NTU, Waddington, Lossiemouth, 8 Squadron *Ermintrude*, 205, 28, 210, 224 Squadrons
Fuselage only, dismantled in four sections	Ex-Narborough, Cosford 8119M, 120, Kinloss Wing, 201, 120, Kinloss Wing.
Currently being reassembled and moved to Bruntingthorpe	Ex-Cosford 8117M, Kinloss Wing, 203, 42, 203, SWDU, ASWDU
Complete	Ex-Finningley 8186M, 203, 42, 206, 203, 42, 201, 206, 201, 220 Squadrons
Complete – undertakes occasional running of engines	Ex-Cosford 8116M, 201, 206, MoA, 205, 203, 206 Squadrons
Complete but derelict	Ex-Cosford 8103M, 201, 120, 206, 203, 206, A&AEE, 206 Squadrons
Complete but partially dismantled for restoration	Ex-Kemble, 203, 120, 201 Sqn. Arrived 22.08.72
Complete – abandoned at western end of runway 11/29	
Complete – abandoned at western end of runway 11/29	
Fuselage section – abandoned at western end of runway 11/29	
Airframe only – abandoned after spares reclamation	
Cockpit section only	Ex-St Austell, St Mawgan 9109M, 8, 204, 205, 37, 38 Squadron. SOC 23.02.88
Complete but abandoned in desert	Ex-35 Squadron
Complete	Ex-35 Squadron
Complete – last 'airworthy' example. Occasionally ground runs.	Ex-35 Squadron
Complete – located above garage	Ex-35 Squadron
Complete	

Bibliography

Ashworth, Chris, *The Shackleton: Avro's Maritime Heavyweight* (Aston Publications, 1990).

Barnes, C.H., *Bristol Aircraft since 1910,* 1st edition (Putnam & Company, 1964).

British Aviation Research Group, *British Military Aircraft Serials and Markings*, 2nd edition (BARG/ Nostalgair/The Aviation Hobby Shop, 1983).

By Command of the Defence Council, *The Malayan Emergency 1948–1960* (Ministry of Defence, 1970).

Chartres, John, *Avro Shackleton* (Ian Allan, 1985).

Ellis, Ken, *Wrecks & Relics*, 24th edition (Crécy Publishing, 2014).

Flintham, Victor, *Air Wars and Aircraft – A Detailed Record of Air Combat, 1945 to the Present* (Arms and Armour Press, 1989).

Flintham, Victor and Thomas, Andrew, *Combat Codes – A Full Explanation and Listing of British, Commonwealth and Allied Air Force Unit Codes since 1938* (Pen & Sword Aviation, 2008).

Franks, Richard A., *Shackleton, Guardian of the Sea Lanes* (Dalrymple and Verdun Publishing, 2005).

Jefford, Wing Commander C.G, MBE, RAF, *RAF Squadrons* (Airlife, 1988).

Jones, Barry, *Avro Shackleton* (The Crowood Press, 2002).

Lake, Deborah, *Growling over the Oceans* (Souvenir Press, 2010).

Lee, Air Chief Marshal Sir David, GBE, CB, *Flight from the Middle East* (Air Historical Branch, 1978).

Lee, Air Chief Marshal Sir David, GBE, CB, *Wings in the Sun – A History of the Royal Air Force in the Mediterranean, 1945–1986* (HMSO, 1989).

Lindsay, Robert and Heeley, Howard, *Duty Carried Out 1957–1971* (Newark Air Museum, 1995).

Lindsay, Robert, Otter, Andy and Heeley, Howard, *Dedication to Duty* (Newark Air Museum, 1999).

Ritchie, Dr Sebastian, *The RAF, Small Wars and Insurgencies: Later Colonial Operations, 1945–1975* (Air Historical Branch, 2011).

Robertson, Bruce, *British Military Aircraft Serials, 1878–1987* (Midland Counties Publications, 1987).

Thetford, Owen, *Aircraft of the Royal Air Force since 1918*, 8th edition (Putnam, 1988).

Trevenen James, A.G., *The Royal Air Force – the Past 30 Years* (McDonald and Jane's Publishers Limited, 1976).

Various editions of *Air Clues* magazine, issued monthly for the Royal Air Force by the Director of Flying Training (MoD).

Various editions of *Flight International* magazine, published weekly by IPC.

Index